Geopolitics and the Post-colonial

D0293840

To Farida

For her generosity of spirit and caring wisdom

Geopolitics and the Post-colonial

Rethinking North–South Relations

David Slater

Blackwell
Publishing

© 2004 by David Slater

BLACKWELL PUBLISHING
350 Main Street, Malden, MA 02148-5020, USA
108 Cowley Road, Oxford OX4 1JF, UK
550 Swanston Street, Carlton, Victoria 3053, Australia

The right of David Slater to be identified as the Author of this Work has been asserted in accordance with the UK Copyright, Designs, and Patents Act 1988.

All rights reserved. No part of this publication may be reproduced, stored in a retrieval system, or transmitted, in any form or by any means, electronic, mechanical, photocopying, recording or otherwise, except as permitted by the UK Copyright, Designs, and Patents Act 1988, without the prior permission of the publisher.

First published 2004 by Blackwell Publishing Ltd

Library of Congress Cataloging-in-Publication Data

Slater, David, 1946-
    Geopolitics and the post-colonial : rethinking North–South relations / David Slater.
        p. cm.
    Includes bibliographical references and index.
    ISBN 0-631-21452-6 (hardback : alk. paper) – ISBN 0-631-21453-4 (pbk. : alk. paper)
1. International relations. 2. Geopolitics. 3. Postcolonialism. 4. World politics–20th century.
5. North and south. I. Title.

JZ1251.S58 2004
327.1'01–dc22   2004008267

A catalogue record for this title is available from the British Library.

Set in 10.5/13 pt Sabon
by Kolam Information Services Pvt. Ltd, Pondicherry, India
Printed and bound in the United Kingdom
by MPG Books Ltd, Bodmin, Cornwall

The publisher's policy is to use permanent paper from mills that operate a sustainable forestry policy, and which has been manufactured from pulp processed using acid-free and elementary chlorine-free practices. Furthermore, the publisher ensures that the text paper and cover board used have met acceptable environmental accreditation standards.

For further information on
Blackwell Publishing, visit our website:
www.blackwellpublishing.com

Contents

Preface

This book has grown out of a long-standing interest in North–South relations and their crucial connection with the power of geopolitical interventions. The analytical emphasis falls on ideas and issues that have been and remain of key significance to this connection. The approach taken is broadly based in post-structuralist and post-colonialist thinking, with associated links to critical development theory, and the contextualization mainly relies on illustrative examples selected from the trajectory of US–Latin American relations. The geopolitical perspective has been influenced by the creative research falling under the rubric of 'critical geopolitics', and my prioritization of a post-colonial analytical sensibility draws its inspiration from a wide range of authors stretching across the North–South divide.

While the post-colonial in its general analytical form is closely associated with the incisive contributions of Said, Spivak and Bhabha, in the specifically Latin American context, the work of Coronil and Mignolo, especially their critique of Occidentalism, has been particularly relevant to my own approach. As part of a post-colonial perspective, I have integrated the ideas and theoretical reflections of Third World scholars and thinkers throughout the nine chapters, and this is especially the case with respect to the chapters on dependency and difference (5 and 6). In this sense, it can be restated that the subaltern not only speak, act and write back – they also form an intrinsic part of the globality of knowledge. To be indifferent to this reality is to help sustain the imperiality of knowledge which a post-colonial perspective seeks to contest and to disrupt. Although it might be argued with some justification that formally we live in a post-colonial world, we are still some way from inhabiting a post-imperial world.

The book is divided into four parts. The first part sets out to establish some of the most relevant guidelines for the conceptual and historical framework. Chapter 1 includes a review of the debates on categories such as First World/Third World, North/South, West/non-West, and does so in a context that seeks to identify and question the limits of Euro-Americanist approaches to power and knowledge. It is argued that power and knowledge are most appropriately viewed as relational, and as not being confined to any one spatial sphere, whether global or national. The chapter concludes by suggesting how we might think about 'the critical' in critical thought. Chapter 2 provides one possible historical perspective on the development of US power from the nineteenth century to the Second World War and does so in the context of aspects of US–Latin American relations. The conceptual focus falls on questions of an emerging imperiality of power, the nature of geopolitical representations of the societies of the Latin South, and the significance of Latin American counter-representations of US expansionism.

The second part of the book discusses two waves of Western theory (modernization and neo-liberalism) and does so in the setting of changing geopolitical realities since the end of the Second World War. Differences and similarities between these two perspectives are identified and considered in the changing context of North–South relations and the impact of the Cold War and its demise. In both cases, emphasis falls on the mutations in theory as well as the dynamic of world politics and their complex interrelations. The third section takes up issues relating to alternative modes of thinking. Chapter 5 re-visits dependency writing as part of a project of critical memory, and in chapter 6 more recent analyses, from the post-modern to the post-colonial, are set in a context of debates on identity, difference and agency. Overall, parts II and III of the book provide a frame for examining what I call post-colonial questions for global times.

These questions are the primary subject of part IV, which includes a rethinking of the geopolitics of the global and re-examines the contemporary importance of imperial power and its resurgence (see chapter 7). Resistance to neo-liberal globalization and the place of oppositional forms of subjectivity, as embodied in the Zapatista uprising and the World Social Forum, are the focus of chapter 8, which leads into a concluding chapter that sets out the case for going beyond imperial knowledge.

In approaching geopolitics through a post-colonial vision, the book offers one possible analytical pathway to a rethinking of North–South relations. At the same time, I hope that it might contribute to a renewed focus on the need to challenge the imperiality of power and knowledge.

Acknowledgements

Ideas and viewpoints that helped to structure my own approach to geopolitical questions and North–South relations have developed out of an interactive process with colleagues and friends. In the Netherlands, I would like to thank the following people, with whom I have had many fruitful exchanges: my good friend and colleague the late Jean Carrière, Cris Kay, Norman Long, and Frans Schuurman. As a result of regular research visits to Latin America I have been able to establish contact with many writers and scholars and I have greatly benefited from intellectual engagement with a wide range of people, including Lucio Kowarick, Pedro Jacobi, the late Milton Santos, and Zander Navarro in Brazil; Fernando Calderón, Roberto Laserna, and Humberto Vargas in Bolivia; Guillermo Labarca, Martín Hopenhayn, and Carlos de Mattos in Chile; Baltazar Caravedo and Eduardo Ballón in Peru, and Pablo González-Casanova and Hugo Zemelman in Mexico. I would also like to mention Sonia Alvarez, Evelina Dagnino, and Arturo Escobar, who kindly invited me to take part in their Latin American Studies Association research group on social movements, and chapter 8 reflects some of the ideas produced by this group.

At Loughborough, I have been fortunate to be in a Geography Department with colleagues whose research is innovative, critical and international in focus. In particular, an interest in globalization, post-coloniality and North–South relations is shared by a number of colleagues (including Morag Bell, Ed Brown, Tracey Skelton, and Peter Taylor) with whom I have had many thought-provoking discussions. Latin American research has also been carried forward by a group of postgraduate students, including Yubirí Aragort, Jon Cloke, and Alberto Cortés Ramos. I would also like to thank Loughborough University for providing the financial support that has enabled me to sustain regular research visits to Latin America and for facilitating a period of study leave which gave me the time to complete the book.

At Blackwell, Angela Cohen, Justin Vaughan, and Simon Alexander have provided positive encouragement and a patient approach to shifting deadlines. Finally, my partner, Farida Sheriff, has given me continuing intellectual and emotional support, stimulating me to rethink many of my initial ideas, while also, after reading through the manuscript, helping me to sharpen and strengthen the entire text.

Part I

Conceptual and Historical Issues

1

For a Post-colonial Geopolitics

'The central point... is that human history is made by human beings, and [s]ince the struggle for control over territory is part of that history, so too is the struggle over historical and social meaning. The task for the critical scholar is not to separate one struggle from the other, but to connect them...'

– Edward Said (2003: 331–2)

Together with the post-1989 dissolution of the Second World, the accelerating tendencies of globalization and the explosive surfacing of a variety of acute social tensions and conflicts, there has also been a resurgence of interest in the state of North–South relations. Already in the early 1990s, it was suggested that the growing gap between First and Third Worlds was raising some of the most acute moral questions of the modern world and becoming a central issue of our times (see Arrighi 1991 and Hösle 1992). This re-assertion of the significance of North–South relations captures one of the world's geopolitical continuities. Thus, in a world frequently portrayed in terms of flows, speed, turbulence and unpredictability, there is another narrative rooted in historical continuity – the recurring stories of poverty, inequality and exclusion – a 'shock of the old'.

For example, global inequalities in income in the twentieth century have increased by more than anything previously experienced, illustrated by the fact that the distance between the incomes of the richest and poorest country was about 3 to 1 in 1820, 35 to 1 in 1950, 44 to 1 in 1973, and 72 to 1 in 1992 (UNDP 2000: 6). Inequalities are also to be symptomatically encountered in the world of cyberspace, where access to the internet displays a familiar geographical distribution, with over 90 per cent of all internet hosts being located in developed countries which account for only 16 per cent of the world's population (see Main 2001: 86–7 and also World Bank 1999: 63).

On poverty and exclusion, the UN Human Development Report for 2002 notes that 2.8 billion people still live on less than US$2 a day (UNDP 2002: 2), while a previous Report stated that in Third World countries more than a billion people lack access to safe water, and more than 2.4 billion people lack adequate sanitation (UNDP 2000). Information on hunger, or the degree of malnutrition in the world, can be found in the FAO World Food Survey for the early 1990s, which concluded that a little over 800 million people (approximately one-sixth of the world's population at that time) suffered from undernourishment in 1990–2 (see Sutcliffe 2001: 48), a figure that had risen slightly by 1996 (Diouf 2002).[1]

Also in the 1990s, although the number of people killed in conflicts between countries had fallen sharply from the 1980s, about 3.6 million people died in wars within states, and the number of refugees and internally displaced persons had increased by 50 per cent (UNDP 2002). Equally if not more disturbing, there is evidence of new forms of slavery in the global economy. Bales (1999), for example, estimates that in the 1990s there were 27 million slaves, of which perhaps 15 to 20 million were working as bonded labour in India, Pakistan, Bangladesh and Nepal; there are more slaves alive today, writes Bales, than all the people stolen from Africa in the time of the transatlantic slave trade.

These figures point to the continuity of inequality, poverty, forms of social and economic exclusion and oppression, and the reality they reflect is frequently discussed in analyses of the North–South divide in an era of globalization. Equally, it needs to be remembered that there have been socio-economic and political improvements as well. For example, whereas in 1900 no country had universal adult suffrage, by the end of the 1990s the majority of the world's countries did; and also, in the last three decades, life expectancy in what the UN refers to as the 'developing countries' increased from 55 years in 1970 to 65 in 1998 (UNDP 2000). On the directly political terrain, and again with the UN's Human Development Report, we are reminded that in the 1980s and 1990s the world made dramatic progress in expanding political freedoms, so that today 140 of the world's nearly 200 countries hold multi-party elections – more than ever before (UNDP 2002: 1).

While it is evident that this kind of description and identification of some basic trends could be taken much further, this is not my intention here. Rather, I want to pose certain questions concerning the categories that are used to order and classify these varied features of what is often referred to as the North–South divide. Specifically, it is important to consider the ways in which the societies of Africa, Asia, Latin America, the Caribbean and the Middle East have been generally represented, and also how the writers and intellectuals of these regions have generated counter-representations of their own realities and also of the relations between North and South or First World and Third World. Many of these aspects will be taken up in more detail in the following chapters, so the purpose of this introductory section is to make an initial exploration of the analytical terrain, and at the same time to clarify my own approach.

Categories in Question

It might be contended that in a fast-moving world, the long-established categories of First World/Third World, West/non-West and more recently North/South are increasingly obsolescent. In global times, they are simply the residual traces of discourses of social change that have been shorn of any effective contemporary meaning. Responding to such an overall orientation, and more specifically to the suggestion that First World/Third World distinctions can be dissolved without any loss of analytical scope, a number of Latin American writers have stressed the need to rethink patterns of global inequality, rather than neglect or deny their continuing significance (see, for example, Hopenhayn 1995 and Richard 1995).[2] While such a response is helpful, it still leaves us with the task of clarifying the meanings, potential relevance and also differentiation of the above categories. First of all, and as a key example, I shall briefly consider the First World/Third World couplet, with particular attention to the latter term.

As is well known, the classification of three worlds of development basically dates from the early 1950s (Pletsch 1981), the dawn of the 'development era' (see chapter 3 below). For Worsley (1979: 102), the Third World was constituted through three interrelated features: non-alignment, poverty and a colonized status. For Jalée (1968), the so-called Third World was no more than the backyard of imperialism, and the countries falling under this rubric were not separate from but very much part of the imperialist system. Further to these initial characterizations, which linked the Third World to non-alignment, poverty, colonialism and imperial exploitation, how has the term been defined since the 1950s and 1960s, and what do these representations tell us?

One significant interpretation can be found in Hobsbawm's (1994) *Age of Extremes*, where the Third World is defined in terms of 'instability' and 'inflammability', leading on to the idea that, before its gradual decomposition and fission in the era of globalization, the Third World differed from the First World in one fundamental respect – 'it formed a *worldwide zone of revolution* – whether just achieved, impending or possible' (Hobsbawm 1994: 434, emphasis added). Hobsbawm goes on to note that very few Third World states of any size went through the period since 1950 without revolution, military coups to suppress, prevent or advance revolution, or some other form of internal armed conflict. In other words, and in the context of the Cold War and super-power rivalry between the US and the Soviet Union, the Third World was a 'zone of war', reflected in the fact that in the period between 1945 and 1983 over 9 million people were killed in East Asia, 3.5 million in Africa, 2.5 million in South Asia, and somewhat over half a million in the Middle East (ibid.).

The profiling of the Third World in relation to war and political instability lent itself to a notion of the Third World as threat, and specifically as threat to Western security. Discussion of a so-called 'Third World threat' in relation to US foreign policy can be found in the early 1970s (Gardner, LaFeber & McCormick 1973: 476), and in the 1980s Kolko (1988) published a book on US geopolitical strategy since the end of the Second World War, entitled *Confronting the Third World*. More recently, in the post-Cold War era, issues of immigration, combating terrorism, and the traffic in illegal drugs, have led some authors to talk of an 'increasing Third World threat' to American interests (David 1992/3: 156; and for other examples see Furedi 1994: 115–16). My point here is to signal that the representation of the Third World as a threat to First World stability dates back to before the end of the Cold War and the dissolution of the Soviet Union and the Second World. Moreover, as will be seen in the next chapter on US–Latin American relations before the Second World War, notions of danger and disorder were regularly deployed to situate the societies of the Latin South.

The connection between the Third World and instability and disorder is frequently treated as if this were an innate, indigenous feature of the Third World itself, but as can be shown this ignores the history of colonial penetration. For example, in his revealing book on famines and the making of the Third World, Davis (2001) stresses the point that what is today called the 'Third World' is the outgrowth of income and wealth inequalities that were 'shaped most decisively in the last quarter of the nineteenth century when the great non-European peasantries were initially integrated into the world economy' (2001: 16). Through invasion and war, disorder and poverty were externally induced, a theme to which I shall return in subsequent chapters.

Staying with Davis, I want to say something about another aspect of the way in which the term Third World can be used to characterize a certain reality within the First World itself, exemplifying a kind of conceptual stretching.

In a dialogue concerning new thinking on Los Angeles, Davis (Davis & Sawhney 2002: 30–1) contends that First and Third Worlds have co-existed across generations and in different places, echoing Sawhney's comment that First and Third Worlds share many of the same economic resources, languages and geographical locations, so that it is becoming increasingly difficult to separate them (ibid.). This kind of interpretation can be linked to Rieff's (1991) text on *Los Angeles: The Capital of the Third World*, wherein the term Third World is applied to spaces inhabited by domestic minorities such as 'women of colour' and to 'underprivileged' ethnic and social groups. While one may want to question the metropolitan presumption that lies behind the designation of a US city as the capital of the 'Third World', it does raise an important question concerning the meaning and fluidity of the Third World category.

Similarly, Deleuze and Guattari (1988: 468–9) stretch the original meaning of the Third World and bring it back to the First World. They suggest, for example, that the capitalist states of the centre not only have external Third Worlds but also internal Third Worlds; these are seen as peripheral zones of underdevelopment inside the centre in which 'masses' of the population are abandoned to erratic activities such as work in the underground economy.

In more specific geographic terms, marginalization or exclusion have been defined not as a 'Third Worldization' but as a 'Brazilianization' of the world. This, for Beck (2000: 51), in his book on globalization, would be a division of the world sparked by the exclusion of those 'without purchasing power', perhaps the future majority of mankind. This suggestion is tied into a dystopian vision of the 'Brazilianization' of Europe in which one encounters societal breakdown (ibid.: 161–3).

In relation to these connected deployments of the Third World category, there are four points I would like to make.

1) The way the Third World is defined in the above texts underlines its association with poverty, exclusion, underdevelopment and multicultural identities. What tends to be underplayed, with the exception of Davis, and to some extent in Deleuze and Guattari, is any significant recognition of the asymmetrical relations intrinsic to colonial and imperial power. Thus while the Third World is made part of the First World through its identification with poverty, exclusion and hybridity, this misses out of account another rather crucial dimension of Third World identity, namely the fact that the Third World has never made the First World an object of colonial expansion, never developed an imperial gaze over the societies of Euro-America. This might be seen as being a transparently obvious point, but the omission of a defining feature of Third World political identity constitutes an absence that needs to be analytically recovered, as I will argued later.

2) While poverty and exclusion or war and conflict are certainly key facets of Third World societies, and, more specifically, while Brazil does face acute problems of social polarization, violence and drug wars, is that the full picture? From the innovative and successful introduction of participatory urban planning in Porto Alegre or from the founding of the World Social Forum in Brazil, could we not envisage another more creative and positive 'Brazilianization' of the world? Equally with the Third World, is it not possible to think of enabling forms of 'Third Worldization', through perhaps an emphasis on solidarity, reciprocity and collective engagement?

3) Although it is necessary to be critical of ideas about the overlapping of First and Third Worlds, especially when they are divorced from the history of imperial power relations, at the same time the profiling of new interpenetrations can encourage us to take seriously the imbrication of the inside and the outside, of the domestic and the foreign. Not only is this significant in terms of questions of immigration and the fortification of frontiers, as in the case of US–Mexico relations (Nevins 2002), but also there are other interconnections such as the resurgence of racism against asylum-seekers taken together with the re-assertion of imperial politics (for example, notions of 'Western superiority' and the posited beneficence of Empire).

4) Finally, with the varied problems attached to the term 'Third World', and more concretely with the collapse of the Berlin Wall, the subsequent dissolution of the Soviet Union and the effective ending of the Second World, with the outlying exceptions of Cuba and North Korea, the geopolitical setting in which a Third World could be located has been

fundamentally transformed. And yet the term lives on, as witnessed in publications such as *Third World Quarterly* and *Third World Resurgence*, and continues to be used in the social science literature (Thomas 1999). My own approach is to employ the Third World category in discussing modernization theory and dependency thinking, whereas in later chapters that relate to the post-1989 era I shall predominantly use the North–South couplet, which has a contemporary origin in the Brandt Report on Survival and International Development (1980), and which at least avoids the problems of a tripartite division of the world in a post-Cold War era.

Another binary division that I have already alluded to and which permeates a wide range of texts is that of the West/non-West distinction. This couplet has been deployed in a way that grounds a primary identity for the West, as the self, and a secondary identity for the non-Western other. Traditionally, 'the West' has been constructed as a model and measure of social progress for the world as a whole. It has been and remains much more a driving idea than a fact of geography. For example, in the late 1970s, one member of the Argentinian junta asserted that 'the West today is a state of the soul, no longer tied to geography', while for another, 'the West is for us a process of development more than a geographical location' (Graziano 1992: 123 and 271). From a very different political position, the Indian writer Ashis Nandy (1992: xi), in his text on the psychology of colonialism, linked notions of the West with modern colonialism and its impact in the Third World, and also noted that the modern West was as much a psychological category as a geographical or temporal entity – 'the West is now everywhere,' he wrote, 'within the West and outside; in structures and in minds' (ibid.). What this line of argument points to is the suggestion that Third World societies have been colonized by a Western imagination that frames and represents their meaning as part of a project of rule, and examples of such a framing can be found in both modernization and neo-liberal discourses, as will be indicated in chapters 3 and 4.

Geographically, the West, the First World and the North are customarily associated with the countries of Western Europe, North America and Australia and New Zealand, with Japan being classified as both First World and of the North, but clearly more of the East than of the West. What is important to note here is that these binary divisions, particularly First World/Third World and West/non-West, are charged categories (see Bell 1994, for example). They are replete with sedimented meanings, while in contrast the North–South distinction, I would argue, is less burdened with those deeply rooted associations of Occidental or First World primacy. Given the fact that all these categories can be justifiably put into question, while at the same time continuing to retain a broad usage, it can be useful to borrow a term from Derrida (1976), whereby these terms are seen as 'under erasure'. In other words, they can be approached as if there is a line running through them, cancelling them out in their old form, but still allowing them to be read. With such a partial erasure, we can be encouraged to continue to reproblematize their meaning, validity, applicability, etc. while keeping an open space for the possibility of new categories.

As a way of moving the argument forward I now want to critically consider the phenomenon of Eurocentrism, or more generally Euro-Americanism. This kind of geopolitical categorization of the world carries within it both an affirmation of Western primacy and a portrayal of the non-West, or the 'developing world' or Third World, as a dependent and ostensibly inferior other. It is important to signal the limitations of this mode of representation as a way of clarifying the analytical ground.

Contextualizing Euro-Americanism

It is possible to specify three constituent elements of Euro-Americanism.[3] First, Euro-Americanist interpretations emphasize what is considered to be the leading civilizational role of the West through referring to some *special* or *primary* feature of its inner socio-economic, political and cultural life. Hence, Max Weber asserted that the West was the 'distinctive seat of economic rationalism' (1978: 480), and that outside Europe there was no evidence of the 'path of rationalization' specific to the Occident (1992: 25). In a similar vein, although within the Marxist tradition, Gramsci in his *Prison Notebooks* stated that European culture was the "*only* historically and concretely universal culture" (Gramsci 1971: 416, emphasis added). In contemporary political theory, the West is frequently portrayed as the primary haven of human rights, enlightened thought, reason and democracy (see, for example, Žižek 1998). In a related manner, and in the domain of philosophy, Western culture has been depicted as the only culture capable of self-critique and reflexive evaluation (see, for example, Castoriadis 1998: 94).

Second, the special or primary feature or essential matrix of attributes that is posited as being uniquely possessed by the West is further regarded within a Euro-Americanist frame as being *internal* or *intrinsic* to

European and American development. This set of attributes is envisaged in a way that assumes the existence of an independent logic and dynamism of Euro-American development. There is no sense of such development being the result of a process of cross-cultural encounter. Not only is there a process of *self*-affirmation, but also a denial of a potentially beneficial association with the non-Western other. This sense of self-affirmation is often associated with a posited superiority which has permeated many discourses, from progress and civilization through to modernization and neo-liberal development (see chapters 2, 3 and 4 below), and has helped fuel the drive to expand and colonize other cultures.

Third, the development of the West, as situated within a Euro-Americanist frame, is held to constitute a *universal* step forward for humanity as a whole. Such a standpoint has been captured in both traditional Marxist views of a progressive succession of modes of production, and in the Rostowian notion of the 'stages of economic growth' (Rostow 1960) with the West offering the non-West a mirror for its future development. The assertion of universality has deep roots and for both Marxist and non-Marxist traditions Hegel was a primary source. In the early part of the nineteenth century, for example, he defined Europe as the principle of the modern world, being synonymous with thought and the universal (Hegel 1967: 212). Such a vision was later re-asserted by Husserl who stated that 'philosophy has constantly to exercise through European man its role of leadership for the whole of mankind' (Husserl 1965: 178).[4]

These three elements – the primary or special, the internally independent, and the universal – form the basis of Euro-Americanist representations, and they tend to go together with negative essentializations of the non-Western other. There is an insistent belief in the key historical and geopolitical significance of the West as the essential motor of progress, civilization, modernity and development. This is coupled with a view of the non-West as passive or recalcitrant recipient, not dissimilar to the Hegelian view of those peoples as being at a 'low level of civilization'. Such a perspective is not without contemporary resonance. For example, in the field of development studies, one can encounter passages such as the one below from an OECD (1996: 6) document:

> In the early 1950s, when large-scale development assistance began, most people outside the developed countries lived as they had always lived, scraping by on the edge of subsistence, with little knowledge of and no

voice in global or national affairs, and little expectation of more than a short life of hard work with slight reward.

This image of stagnation, lack of knowledge and political participation, pervasive hardship, and negative sameness contrasts with the vibrant reality of industrialization already under way in many parts of Latin America at that time, especially in Brazil, Argentina and Mexico, of heterogeneous currents of nationalism across Asia, Africa and Latin America and of early, albeit sometimes precarious, forms of democratic government in Latin America, established well before the early 1950s. The above passage, which represents one kind of erasure of history, is not unsymptomatic, and can be related to those interpretations which stigmatize the developing countries or 'impoverished countries' for their own ascribed lack of improvement. Giddens (2000: 129), for example, in his chapter on taking globalization seriously, writes that most of the problems that impede the economic development of the 'impoverished countries' are not to be attributed to the global economy itself, nor should they be linked to the self-seeking behaviour of the richer nations; rather 'they lie mainly in the societies themselves – in authoritarian government, corruption, conflict, over-regulation and the low level of emancipation of women'. While these phenomena are not unknown in developed countries as well, Western narratives will tend to treat the social and political problems of the West as specific and relatively separate. They will not be combined to call into question Western society as a whole (see Lazreg 2002).

Overall, one of the determining features of Euro-Americanism concerns the emphasis given to the universalist power of Western reason, thought and reflection. This underscoring of the thinking, reasoning subject goes together with a general avoidance of the importance of war and violence as a background to this posited Western supremacy. The Argentinian philosopher Enrique Dussel (1998) takes issue with the separation of thought from conquest and reminds the reader that 'I conquer' must be given historical and ontological priority over what is considered to be the founding Cartesian *cogito ergo sum* ('I think, therefore I am').

For Dussel, conquest means to take possession of the land and the people of a territory, so that any subsequent formulation of thought and truth must already presume a territorialization based on a self/ other split which can only be fully grasped in the frame of conquest. In this context, modernity, for Dussel, began in 1492 with the invasion of the 'New World', so that the precept 'I conquer' not only preceded but also constituted the practical foundation of 'I think'. It provided a context in which a crucial relation began to emerge between geopolitical power and the territorialization of thought (for a related discussion see Spanos 2000).

Dussel's critique is generally aimed at Eurocentrism, which brings us to the question of the difference that 'America' makes. Why is it necessary to distinguish Euro-Americanism from Eurocentrism? I want to suggest that the term Eurocentrism is often used in a way that implies that America can be incorporated within this category, and that presumably the similarities between Europe and America outweigh the differences. This is a problematic stance since it conceals the specificities of the United States as today's lone superpower, and it also leaves out of account historical and geopolitical differences that are rather crucial to the whole discussion of West/non-West or North–South relations. Moreover, since the role played by the United States in impacting on Third World countries is an important theme of my analysis, it is necessary to begin with an analytical outline that places this issue on the agenda.

Along one analytical route, the differences between the United States and Europe have been considered in the context of 'American exceptionalism' (Bell 1991).[5] This topic has been widely covered in the literature, in some cases in relation to US foreign policy (Agnew 1983) and more recently in the context of continuing US global power (for example, Guyatt 2000). My concern here is rather to focus on what I consider to be three specificities of the United States which are particularly pertinent to a consideration of the relations between the US and the societies of the global South, and which are important for our understanding of Euro-Americanism.

The first difference that needs to be highlighted concerns the nature of US imperial power. Unlike other Western nations, the imperiality of US power emerged out of post-colonial roots. In other words, in the case of the United States a project of Empire emerged out of an initial anti-colonial struggle for independence from British rule in the late eighteenth century. This gives the United States a paradoxical identity of being a 'post-colonial imperial power' – although as will be argued in chapter 2, the key emphasis needs to fall on the imperial, and moreover the 'post-colonial' element of identity did not preclude the holding of a territorial colony, as with the example of the Philippines or semi-protectorates (such as Cuba from 1903 until 1934). There are two features of this contradictory identity that can be identified.

a) In the first place, and in the context of the geopolitics of US interventionism, one finds juxtaposed an affirmation of the legitimacy of the self-determination of peoples, emanating from its own origin, with a belief in the global destiny of the United States. Historically, the contradiction between a belief in the rights of people to decide their own fate and a belief in the geopolitical predestination of 'America' has been officially 'resolved' through the invocation, especially present since the beginning of the twentieth century, of a democratic mission that brings together the national and international spheres. To transcend the contradiction between an identity based on the self-determination of nations and another rooted in Empire, a horizon is provided for other peoples who are encouraged to choose freedom through the democratic way, thereby integrating their own struggles within an 'American' vision. Such an invocation to embrace the American path to freedom and democracy is still very much with us, as I shall show in later chapters.

b) Secondly, the primacy of self-determination provides a key to explaining the dichotomy frequently present in US interventions where a split is made between the governed (the people) and the governors (the rulers). Given the historic differentiation of the New World from the Old, and the support for anti-colonial struggles, perceived threats to US security have been accompanied by a separation between oppressed people and tyrannical rulers. In the enduring example of US hostility towards the Cuban Revolution, a strong distinction has been made between the Cuban people who are portrayed as being oppressed by their communist rulers and the Castro regime. For instance, the Helms–Burton Act of 1996 specifically argues that measures are needed to restore the values of freedom and democracy and above all the sovereign and national right of self-determination to the Cuban people.[6] The geopolitical representation at work here can be described in terms of the assumed right to be able to designate the political future for a people whose sovereignty is envisaged as being usurped by an ostensibly tyrannical regime. As will be seen in chapter 2, such a mode of imperial representation is anchored in the history of US–Cuban relations; and, as we will see in chapter 7, continues through into the twenty-first century.

The second specificity of the United States that is relevant to our theme, relates to the particular territorial formation of the United States, a formation that was intrinsically grounded in war and the expansion of a 'civilizing' frontier. In this formation there were encounters with three significant others: the indigenous peoples of North America (the

'Indian'), the Hispanic and Indian population of Mexico, confronted in the US–Mexico War of 1846–8, and the African American in the context of slavery and its abolition after the 1860s Civil War. In the founding example of the decimation of the native peoples of the continent, white America's violent encounter with its Indian other came to form a pervasive element of the nation's collective memory. It not only figured in the production of films about how the 'West was won', but also found expression in twentieth-century warfare, so that, for example, during the Vietnam War, US troops described Vietnam as 'Indian country' (Slotkin 1998: 3, and for a similar example in the case of intervention in Bosnia, see Campbell 1999: 237).

Notwithstanding the historical and cultural differences between them, these founding three encounters with internal others were marked by forms of subordinating representation and mechanisms of power that prefigured subsequent US encounters with Third World others. Before the United States became a global power, these encounters provided an original reservoir of imperial experience that was not irrelevant to many of the interventions pursued by the United States in the twentieth century. In comparison to the colonial powers of Western Europe,[7] in the case of the United States, the internal territorial constitution of the nation-state was striated by a series of violent encounters with other peoples that took place on its own soil and intimately moulded its evolving sense of Empire and mission, perhaps nowhere more visibly reflected than in the significance given to notions of 'the frontier', a theme taken up in chapter 2.

The third relevant specificity of the United States relates to the way its leading political figures have portrayed their country as the original haven of a 'New World'. From the Monroe Doctrine of 1823 and Thomas Jefferson's twin notion of 'America' having a 'hemisphere to itself' while being an 'Empire for Liberty', through to the Roosevelt Corollary of 1904 and the Rio Pact of 1947, the United States has staked out for itself an original heartland separate from the 'Old World' of Europe. This demarcation of geopolitical domains, or the establishment in the western hemisphere of a 'grand area' of geostrategy, constituted what can be referred to as the first phase of a US strategy of containment. This first phase, which will be discussed in the second chapter, was characterized by a strategy for the establishment of US hegemony in the Americas, and the setting of limits for European influence. The second phase of containment, which began with the Cold War and the rivalry between the superpowers, saw the United States, as a global power, developing a strategy of containment for what was perceived to be the Soviet-led communist threat to the 'free world' of the Western nations. This classic phase of containment was played out on the global stage from the late 1940s to 1989, and is considered in chapter 3. The final phase of containment, which began in the early 1990s, relates to the specific targeting of what are defined as 'rogue states' such as North Korea and Iraq – states which are considered to either harbour terrorist groups or possess weapons of mass destruction or both. With the invasion of Iraq in 2003, this 1990s phase of containment has been replaced by a strategy of pre-emptive attack, and a much more aggressive geopolitical doctrine for re-asserting and spreading US power, a topic to which I shall return in chapter 7. Overall, the evolution of these three interconnected phases of containment thus provides the third specificity of the United States within the West.

In sum, the three specificities outlined above constitute key differences between the United States and the rest of the West. They are vital to a fuller appreciation of the particularities of the projection of US power, and especially in the context of US–Third World relations. Furthermore, in any discussion of the difference between Eurocentrism and Euro-Americanism, these three factors provide one possible basis for understanding the contrasts as well as the commonalities within the West. There is one further important difference within the West that needs to be mentioned, a difference that also has had implications for the coloniality of power.

Frequently, in the way the West is designated – especially, for example, in relation to questions of modernity and globalization – there is a tendency to equate the West and more specifically Western Europe with the countries of Britain, France and Germany. In this often implicit specification, what is sometimes missing is an appreciation of the historical differentiation of Western colonialism, and in particular the specificity of the earlier examples of Spanish and Portuguese colonialism in Latin America. The critique of 'occidentalism', as initiated by Coronil (1996) and Mignolo (1995), has created a highly relevant opening in this regard, since their point of departure is the colonial relation as it emerged and evolved in South and Meso-America. They both argue against an occidentalism that separates the world into bounded units, that disaggregates the relational histories of colonizer and colonized, turns differences into a hierarchy of self and other and helps reproduce asymmetrical power relations. In addition, they usefully underscore an important difference of the post-colonial, namely the earlier Independence of the Spanish and Portuguese colonies of the Americas, so that the Latin

American post-colonial as an historical periodization has its own specificity *vis-à-vis* that of Africa and Asia.

How, then, might we approach the post-colonial in a context of geopolitical representation, and how might we conceptually locate power relations which are intrinsic to the questions at hand?

Power, Geopolitics and the Post-colonial

Let us begin with an argument outlined by one of today's most innovative social theorists. Bauman (1999), in a discussion of power, politics and the territorial, makes a strong case for separating power from politics. He maintains that a key characteristic of our times is the fact that true power, which is able to determine the extent of practical choices, flows, and because of its ever less constrained mobility it is exterritorial. For Bauman, all existing political institutions remain strongly local; and, crucially, the heart of the contemporary crisis of the political process is the absence of an agency effective enough to legitimate and promote a cohesive agenda of choices. Today the principal agenda-setters are 'market pressures' which are replacing political legislation, and while geographical space remains the home of politics, capital and information inhabit cyberspace in which physical space is cancelled or neutralized (Bauman 1999: 120). Power, in this reading, *was* territorial. The era of space was the time when borders could be made impermeable; land was a shelter and a hideout, and the powers-that-be from which one wished to escape stopped at the borders. Now, for Bauman, post-September 11, this is all over, for no one can cut themselves off from the rest of the world. Places no longer protect, and strength and weakness, threat and security have become '*extraterritorial* (and diffuse) *issues that evade territorial* (and focused) *solutions*' (Bauman 2002: 88–9, emphasis in the original).

There are a number of points here which are suggestive for and germane to the analytical pathway I seek to explore.

First, one might remind ourselves of the Foucauldian-based distinction between the 'power-over' and the 'power-to'. The former is more associated with a state of domination, whereby, for example, an agent is able to exert an impinging and moulding influence over another agent, whereas the 'power-to' can be associated with the ability to resist the influence of another, and can be seen in the emergence of collective actions and the impact of social movements. In both kinds of power, there is surely no reason to exclude *a priori* any domain or space of effects. With power-over, the forces of globalization (for example, the role of transnational corporations) may be seen as part of the space of flows that Bauman emphasizes, but that kind of power-over requires a connection with state power, so that it can be more effectively internalized and is in that sense both territorial and extraterritorial in Bauman's terminology. Equally, historically and presently, state structures in many parts of the world continue to deploy their own rooted capacity to influence, sometimes through coercion, sometimes through persuasion, the actions of their territorialized subjects.

What I am arguing therefore is that while in the era of globalization, power may well be seen to flow more than in earlier periods due to the increased effectiveness of communication networks and a generalized global interconnectedness, this need not lead us into ignoring the persist-ence of older forms of the power-over which exist within the territorial ambits of nation-states (considered in chapter 4). Also, power-to, ex-pressed in the form of collective mobilizations, has found two different although interrelated instances of expression. First, within the bound-aries of nation-states, as for example with a range of indigenous movements in Latin America, including the Zapatistas in Mexico, struggles have emerged which, while developing transnational links, are essentially rooted in demands to secure new rights and recognition within specific territories. And, second, in the context of, for example, the World Social Forum founded in Porto Alegre in Brazil, movements have come together across national spaces to forge new kinds of associa-tive power which are both transnational and national (both examples will be examined in chapter 8). Similarly, in the global protests against the war on Iraq, a power of refusal and opposition is also a part of an alternative 'flow of power', especially linked through cyberspace and the globality of the televised image, but also anchored within national spaces. In this sense then 'power-to' cannot be justifiably contained at the extraterritorial level; it flows both within and across national boundaries.

Second, in this complex field of analysis, it is worth recalling an observation made by Foucault, who suggested that power is constructed and functions on the basis of particular powers and a myriad of issues that stretch from the macro-structures of the economy through the domain of the state to social institutions such as the school, the hospital and the asylum (Foucault 1980a: 188). A whole history of spaces remains to be written, Foucault argues in a well-known passage; and this would also be the history of powers, from the 'great strategies of geo-politics to the little tactics of the habitat' (Foucault 1980a: 149). It is exactly this sense of openness and plurality to the analysis of power that I would like to positively underline, in contrast to a possible analytical separation of power from territoriality and politics.

What also needs to be taken into account, and this is my third point from the Bauman passage, is the linkage between the relations of power and the modes of knowledge that give power its potential for effectivity. As Foucault (1979: 27) expressed the point, there is no power relation without the correlative constitution of a field of knowledge. Knowledge and power are intertwined, and the way knowledges are represented and deployed is crucial to that interrelation.

Furthermore, and very much linked to issues of representation, a post-colonial perspective would question the geographies of reference for self and other, and their interrelation or intersubjectivity. What is missing in both Bauman and Foucault is a sense of the difference that colonialism or Empire makes to the ways in which power, politics and knowledge combine and work out their effects on the landscape of social change. Spivak (1999), in her work on the post-colonial, which includes a critique of Foucault and Deleuze, has reminded us of the 'sanctioned ignorance' and occlusion of the colonial and imperial moment in Western post-structuralist thinking. The ways in which non-Western others have been and continue to be represented is reflected in a range of subordinating forms of classification, surveillance, negation, appropriation and debasement, as contrasted to a positive self-affirmation of Western identity (Spurr 1993).[8] These forms of representation, incisively analysed by Said (1978 and 1993), find expression within the frame of North–South relations post-1989, as Doty (1996) has shown, and their production is crucial to the sustainability of particular relations of power and subordination.

As has been outlined above, Euro-Americanism exemplifies many of the problems associated with the depiction and representation of non-Western societies, and the elements I mentioned in that section could be considered in a geopolitical setting as having three interwoven components – representations of:

a) the other, e.g. the Third World;
b) the self, e.g. the First World and,
c) the interrelations between self and other, e.g. First World/Third World relations.

Frequently critiques of the geopolitics of representation focus on (a) and (c) so that in the example of dependency perspectives (discussed in chapter 5) the critical assessment of modernization theory focused on the inadequate portrayal of Third World reality (a) and the overly sanguine depiction of First World–Third World relations (c), whereas the image drawn of the First World self was subjected to much less critical scrutiny, even though, it might be suggested, that representation was quite vital to the functioning of the theory, as also is the case with the neo-liberal discourse of development (see chapter 4). These three inter-secting components need to be borne in mind in the development of any critique of the state of North–South relations and they can be seen as an important part of any post-colonial perspective. How might such a perspective be initially specified? I want to outline five elements, to which I shall return in chapter 6.

1. As an analytical mode, as distinct from a historical periodization, the post-colonial seeks to question Western discourses of, for example, progress, civilization, modernization, development and democracy, by making connections with the continuing relevance of *invasive* colonial and imperial power that these discourses tend to evade.

2. The post-colonial can be employed to highlight the mutually constitutive role played by colonizer and colonized, or globalizers and globalized.

3. The post-colonial as a critical mode of enquiry can be used to pose a series of questions concerning the location and differential impact of the *agents* of knowledge. Not only does a post-colonial perspec-tive consider the thematic silences present in influential Western discourses, it also challenges the pervasive tendency to ignore the contributions of African, Asian and Latin American intellectuals and their *counter*-representations of West/non-West relations.

4. Fourth, as a mode of analysis, the post-colonial seeks to give key attention to the 'centrality of the periphery', to foreground the peripheral case since, as the Uruguayan writer Eduardo Galeano (1983: 184) once put it, it is 'in the outskirts of the world... that the system reveals its true face'.

5. Fifth, the post-colonial in terms of the way I interpret it in this text carries with it an ethico-political positionality that seeks to oppose the coloniality and imperiality of power and re-assert the salience of

autonomy and popular resistance to Western penetrations. This is an issue to which I shall return in subsequent chapters.

As part of any development of critical thought, a post-colonial perspective needs to be open to a range of conceptual and thematic routes. In my own case I shall pursue an analytical pathway that seeks to combine a post-colonial perspective with a post-structuralist mode of analysis in which questions of difference, representation and agency are foregrounded in a way which is much more post-Marxist than ex-Marxist. In contrast to Robert Young (2001: 6–7), who in his book on post-colonialism contends that the historical role of Marxism provides a fundamental framework for post-colonial thinking, and with Spivak (1999), who tends to reduce Marxist thought to the economic, I shall take up ideas from a more Gramscian-based Marxism, which, as I suggest in chapter 6, provides a range of concepts such as hegemony and collective wills which still have a vital contemporary relevance.

Hegemony in a Gramscian sense goes beyond domination and is defined as the combination of consent and coercion whereby powers of persuasion and moral and cultural leadership are integral to the concept. With the concept of collective wills, mobilizations and collective actions are not predicated on class belonging but are more part of a fluid movement of opposition to the given disposition of power relations, a movement that transcends class boundaries, even though in the 'last instance' Gramsci himself did not abandon the centrality of class. This notion of a collective will can be compared to Foucault's (1980b: 96) suggestion that in the field of power relations there are always points of resistance (the Gramscian collective will) which are spread over time and space at varying densities, at times mobilizing groups or individuals in specific ways. Clearly these sorts of concepts need to be continually rethought – as Laclau and Mouffe (2001) have done in their own work which breaks away from the centrality of class analysis so characteristic of traditional Marxist thinking.

One way of pursuing the link between politics, representation and power is to focus on projects of hegemony, as with neo-liberalism, discussed in chapter 4, and counter-hegemony, as illustrated in the emergence and development of collective wills, exemplified in the new struggles or points of resistance against neo-liberal globalization and more recently against the re-assertion of US imperialism. (See chapters 7 and 8.) Above all, what needs to be transcended are those kinds of analysis that either (a) treat Marxist categories as 'terminal abstractions' never to be reproblematized or replaced, or (b) summarily dismiss Marxist thought in its entirety as being obsolescent, the expression of a faded utopia. My opening position is that Marxist thought, especially in its Gramscian variant, still has relevance today, but that relevance has to be continually rethought in a critical manner, as part of a wider body of social and political theory. Clearly there are many issues here that require more elaboration, and I shall return to these in chapter 6.

Finally in this chapter I need to clarify my approach to geopolitics.

An initial line of argument would be that any theoretical take on geopolitics needs to be explicitly informed by some specification of what is meant by politics. As I have indicated above, in my brief comments on Bauman and Foucault, it would seem helpful to examine politics and power relations as mutually constitutive. In this context, conflict, division and antagonism are part and parcel of the unfolding of power relations and political processes. Mouffe (1995 and 2000) has explained this argument by suggesting a useful distinction between 'the political' and politics. For Mouffe, 'the political' relates to the antagonistic dimension that is inherent in all human society; an antagonism can take many different forms and can be located in diverse social relations, whereas, in contrast, 'politics' can be taken to refer to the ensemble of practices, discourses and institutions that seek to establish a certain order and to organize social life in conditions that are always potentially subject to conflict because they are affected by the dimension of 'the political'. In this vein, politics can be seen as the attempted pacification of the political, or the installation of order in a given society. Depoliticization is, as writers such as Honig (1993) and Rancière (1995) have eloquently argued, the most established task of politics.

The reference to the political does not entail a marginalization of the formal sphere of politics; rather it calls for a distinction between two spheres that implicate and involve each other. Politics has its own public space – it is a field of exchanges between political parties, of governmental affairs, of elections and representations, and in general of the types of activity, practices and procedures that take place in the institutionalized arena of the political system. The political, in contrast, can be more effectively regarded as a type of conflictual relation that can develop in any area of the social; it is a living movement, a kind of 'magma of conflicting wills', as Arditi (1994: 21) puts it. It is mobile and ubiquitous, going beyond but also subverting the institutional settings and moorings of politics. Politics can be thought of as the institutionalization of an order that is designed to overcome or at least confine the threatening conflicts of the political. But 'order' or 'governance' are always a series of regulative and sedimented procedures, practices, codes and categories that can never be absolutely maintained, since the political – the possibilities of subversion, questioning, opposition, refusal and resistance – can never be fully overcome or vanquished. The interventions that constitute a reactivation of the instability that 'order' seeks to pacify, reflect the inseparability of politics and the political. The political is always that irremovable inner periphery at the heart of politics.

One question that needs to be posed at this juncture concerns the potential relevance of the spatial for any delineation of politics and the political. What difference, for example, does the prefix 'geo-' make to the above argument? One possible answer involves an examination of the relations between nation-states, located in a global structure of such units, which have been traditionally regarded as the building blocks of geopolitics, or as 'containers' that now seem to be increasingly perforated with their powers already subject to significant corrosion (see Agnew 1999, Held 1995 and Taylor 1995). What this issue also raises is the need to distinguish the different spatial levels or spheres within which and across which politics and the political take on their varied meanings. It is possible to suggest the existence of six such spheres ranging from the global and the supra-national through the national to the regional, local and communal levels, the last three being contained within the sphere of the nation-state. My own argument would be to regard the nation-state as a geopolitical pivot, subject to pressures from above and below.

Thus, it can be argued that whereas within the frame of global politics there is more interdependence, the pace of cultural communication, military delivery, disease transmission and so on have accelerated, and that while global issues of refugees, ecology, arms control, organized crime and terrorism have become more intense, *nevertheless* the territorial state remains the most visible and organized site of political action in the world (Connolly 2001). It is a crucial *crossroads* for politics, the political and the spatial. But are all nation-states geopolitically positioned in the same way? Clearly they are not; and what needs stressing in the context of a post-colonial perspective on North–South relations is the difference that both coloniality and imperiality have made. Making this connection is also part of the geopolitical. How?

Customarily, the analysis of the relations between politics and the political is worked out within the conceptual confines of an implicitly

Western territorial state. There is an assumption of a pre-given territorial integrity and impermeability.[9] But in the situation of peripheral polities, the historical realities of external power and its effects within those polities are much more difficult to ignore. What this contrast points to is the lack of equality in the full recognition of the territorial integrity of nation-states. For the societies of Latin America, Africa and Asia, the principles governing the constitution of their mode of political being were deeply structured by external penetration, by the invasiveness of foreign powers. The framing of time and the ordering of space followed an externally imposed logic, the effects of which still resonate in the post-colonial period. The struggles to recover an autochthonous narrative of time, to counter a colonialist rule of memory, and to rediscover an indigenous amalgam of meanings for the territory of the nation have formed a primary part of post-Independence politics. In what were referred to as 'wars of national liberation', the struggle to breathe new life into the time–space nexus of independence lay at the core of the anti-imperialist movement. This then is one modality of the geopolitical, of a transformative rupture, where anti-colonial movements were the disrupting and destabilizing currents able to challenge and eventually bring to an end the colonial appropriation of national space.

But within the bounded territories of nation-states there is another modality of the geopolitical. Across a broad array of societies of the South, movements have emerged which challenge the established territorial orderings of the state. In some instances, such social movements have been rooted in ethnic identities, as has been the case in the post-1994 Zapatista uprising in the Chiapas region of Mexico, while in other cases as in Bolivia, Colombia and Peru, the geopolitical has been partly associated with ethnic-regional mobilizations but also with broader coalitions to restructure and decentralize centralized state power. Here the challenge to the territorial ordering of the central state has assumed a close connection with the notion of extending the territoriality of democratic politics through a decentralization of the state (Slater 2002). In this example, the geopolitical as I have defined it could be also thought of as a *counter*-geopolitics, where an alternative indigenous memory of territory is deployed as part of the ideological struggle against a centralized and mono-cultural state. This is clearly to be seen in the case of the Zapatista uprising in Mexico, discussed in chapter 8.

There are many themes that could be taken up for further consideration. However, at this stage, I want to simply indicate how both geopolitics and its counter, the geopolitical, are best approached as inhabiting a variety of spheres, both international and national, both inside and outside. The 'geo-' in politics and the political needs to be set free from any pre-given anchorage in any spatial level. At the same time, as Bonura (1998) has suggested, critical geopolitical analysis can be made more effective if questions of cultural representation are connected to the power of spatial politics. In my own approach I want to include some suggestions on what it might mean to think critically, in a geopolitical context. There is already an excellent literature in the area of 'critical geopolitics',[10] and in my own case, and as already noted in the preface, I shall develop a perspective that revolves around rethinking the geopolitics of North–South relations in historical perspective, and with specific regard to the relations between the United States and the societies of the South, especially, although not exclusively, the societies of Latin America. But how might we think the 'critical' in critical geographical analysis or more generally the critical in critical thought itself?

On Thinking Critically

The relevance of a critique can be seen in terms of the challenging of what is. It can be envisaged as stressing the importance of investigating the clash of ideas and the durability of difference. It also connects to the impossibility of a political world without adversaries, the myth of a world settled around a ruling consensus. The political does not have a terminal point and, equally, with the critique of what is and what has been, it is necessary to search for what might be. Here I want to propose six possible elements of thinking critically.

1. Analysing presence and absence

This element can be seen in relation to both the themes and concepts of analysis and the agents of knowledge. For example, as part of an appraisal of Western theories of modernization and neo-liberalism, identifying the relative absence of a critical consideration of the history of Western societies and specifically the role therein of colonialism and Empire can form part of the development of critical knowledge. By specifying and analysing the significance of the absence or silence, an alternative vision can be developed. Crucial to this absence is the failure to appreciate the pivotal significance of Western *invasiveness, penetration* and *intervention*. The coloniality and imperiality of power are rooted in the will and capacity to invade and penetrate – the imperial, as contrasted to the colonial, not necessarily requiring the possession of territory. Also, the coloniality and imperiality of power can be used to raise the issue of the imbrication of inside and outside in the sense of tracing the domestic and foreign implications of the colonial and imperial moments. Clearly, Empire, for example, is not a phenomenon that only resides in the international domain; it also affects the domestic terrain, as will be seen in subsequent chapters.

In a similar vein, but in relation to the subjects or agents of knowledge, pointing to the absence of other voices of analysis based in the South constitutes a part of the same critical project of opening up a different kind of interpretive agenda. The exclusion or subordinating inclusion of the intellectual other can be seen as part of the overall politics of occidental privilege. Signalling such an absence and indicating its significance does not have to lead into implicitly underwriting an uncritical reading of the intellectual South. Rather, it is both to question those texts that make the intellectual South invisible and to open up and amplify the analytical terrain – making the absence critically present.[11] Absence, in the way I define it here, therefore, has a duality; it is thematic and conceptual, and also present in relation to the differential exclusion and inclusion of the agents of knowledge.

2. Representing theoretical perspectives

Being critical must also connect to the way different theoretical interpretations are viewed and represented; to what extent, for example, are the *differences* and *commonalities* within given systems of thought such as Marxism identified? The difference for instance between Marx and Gramsci (alluded to above) is significant for the way politics and the political are envisaged, and this difference should alert us to the disadvantage of essentializing a given theoretical tradition. Similarly with post-modernism, it is important to be aware of the different currents within it, as also with post-colonial perspectives (chapter 6). Also with neo-liberalism, while the main signifiers of the perspective need to be specified, equally (as I show in chapter 4) there are important conceptual shifts within the neo-liberal frame which relate to changing geopolitical circumstances.

3. Power/knowledge relations

As was noted above, it is not advisable to separate power from know-ledge, nor power from the politics of discursive representation, including the varied salience of cultural imaginations. Specifically, in the setting of this study, I want to underscore the relevance of connecting the power of a shifting geopolitical landscape, most obviously seen in relation to watershed events such as the collapse of the Berlin Wall in 1989 or the attacks on the United States on September 11, 2001, with the dynamic of conceptual and thematic interpretation. Being critical means not only to be prepared to pursue these connections but also and essentially to challenge the official discourse of world politics, both in relation to the meaning of specific events such as September 11 but also in terms of the underlying markers of interpretation and the geopolitics of strategic action. Furthermore, even though this might be thought to be already understood, social processes in general need to be subjected to critique through the identification of winners and losers, of costs and benefits, so processes such as globalization or modernization are not implicitly en-visaged as being beneficial to all concerned. In addition, and with refer-ence to the difference that agents of knowledge make, the counter-representations of these processes which have been developed in the global South need to be critically included as part of the broadening of our global understanding.

4. Openness to difference and the dynamics of thought

This is an element that is not always easy to pin down, but it relates to being willing to consider different visions while not abandoning a given set of analytical principles which give any perspective its potential ex-planatory coherence, and protect it from a vapid eclecticism. It means being open to learn from opposed interpretations and accepting both the reality of the continuing diversity of knowledge, and the dynamic and contradictions that affect the individual trajectories of specific authors. In the case of Marx, for example, it is clear that there are important shifts of thinking in relation to the way he interpreted non-Western societies, so that in later years his writing on colonialism in India had a closer association with twentieth-century writing on dependency than did his earlier writing, a point that was sometimes passed over.[12] Also, as another rather different example, in the domain of post-modern thinking, Baudrillard's framing of the Third World oscillates from negative essentialization to critical engagement, and taking one text would not do justice to his representation of the Third World (see chapter 6). We always need to leave space open for ambiguity, contradiction, change, complexity and that ever-present precariousness of thought.

5. Deconstruction and reconstruction

Critiques constitute interventions that challenge and renovate the way we think. They are also part of a 'genealogy of knowledge', seen in part as the recovery of hidden interpretations. Critique in the form of genealogy is both rupture and renovation; it foregrounds the embryonic elements of other ways of thinking and is a counter to the cynical reason of one persuasion of post-modern thought where political passivity and feigned omniscience hinder the development of an oppositional consciousness. The questioning of statements and texts through, for instance, tracing the persistence of absences and the repetition of presences can lead to the emergence of alternative readings, of reconstructions that lend us new visions and conceptual pathways. This is also germane to the rereading of intellectual currents that have become less fashionable. Hence, for example, a rereading of dependency perspectives can help in the recovery of ideas that still retain a current applicability.

6. Reflexivity and the geopolitics of knowledge

Finally, a key part of thinking critically involves the situatedness of knowledge, the geographies of reference and the positionality of the writer. From the earliest feminist critiques of political theory, it was suggested that the individual in analyses of political thought was always implicitly a male individual and that masculinity was taken as the norm. The invisibility of women was made visible by feminist writers and the question of who is speaking and whose voice is being heard was situated in a specifically gendered way. Similarly a post-colonial intervention would question an enquiry into a Third World issue where the Third World intellectual is made invisible. The geopolitics of knowledge can help us raise questions that go beyond the thematic and conceptual contours of writing and focus more on the spatial contextualization of analysis itself. This contextualization has frequently been tied to a Euro-Americanist frame, and generalizations have been made which are rooted in the implicit universality of the West. In this text I shall question such implicit universality.

Having outlined six elements of what critical thinking can be about, we are in a position to move on in our consideration of the connections that exist between and through power, politics, representation and space. These connections will be pursued in the following chapters, both historically and geopolitically, being one possible chronicle of the changing state of North–South relations.

2

Emerging Empire and the Civilizing Powers of Intervention

'It is also my wish and expectation that the commissioners may be received in a manner due to the honored and authorized representatives of the American Republic, . . . as bearers of the good will, the protection and the richest blessings of a liberating rather than a conquering nation.'
                                    – William McKinley, 1899 (in Richardson 1905)

'One must demonstrate two useful truths to our America: the crude, uneven and decadent character of the United States, and the continuous existence there of all the violence, discord, immorality and disorder blamed upon the peoples of Spanish America.'
                                    – José Martí, 1894 (in Holden & Zolov 2000: 62)

A Conceptual Opening

A little over a decade ago, Edward Said (1989: 214), in a commentary on a particular current within anthropology, wrote that in so many of the discussions of identity and difference, there is a striking absence of any critical analysis of American imperial intervention. For Said, the imperial contest is a cultural fact of enormous political as well as interpretative significance, since it is this specific historical encounter that acts as a defining horizon for concepts of 'identity', 'otherness' and 'difference'. More generally, and until recently, the absence of accounts of 'American

Empire' in social and cultural studies and its significance for both domestic and international questions has been associated with the literature on 'American exceptionalism' which has tended to separate the republican, liberal-democratic United States from the older world of European colonialism and imperialism. Even in thoughtful analyses of the exceptionalist tradition, as provided for example by the historian Dorothy Ross (1992), the theme of Empire is conspicuously absent. In parallel domains, other examples of the absence of Empire have been found, as in American studies, as shown by Kaplan and Pease (1993). But what exactly do the terms 'Empire' and 'imperialism' signify in the context of the United States?

One way of approaching this question is to refer to Hardt and Negri's recent and influential text on *Empire*. Hardt and Negri (2000: xiv–xv) define Empire in relation to a notion of open, expanding frontiers, where power is distributed in networks. Originating with the founders of the United States, 'the imperial idea has survived and matured throughout the history of the United States constitution and has emerged now on a global scale in its fully realized form' (ibid.). For Hardt and Negri, Empire has no limits, no territorial boundaries limit its reign, and as a concept, Empire presents itself not as conquest but as an order that suspends history, that is outside history or at the end of history. They go on to argue that the fundamental principle of Empire is that its power has no actual and localizable centre and is distributed through networks, which does not mean that the US government and territory are no different from any other – the US does occupy a privileged position, but 'as the powers and boundaries of nation-states decline, ... differences between national territories become increasingly relative' (Hardt & Negri 2000: 384).

An immediate difficulty with Hardt and Negri's interpretation is their idea that the new system of Empire has its roots in the US constitution, which they see as being imperial rather than imperialist, as open and inclusive, so that the policies of Theodore Roosevelt, with his 'international police power' of 1904, represent an old-style European imperialism, whereas Woodrow Wilson and his League of Nations foreshadows today's regime, in which the sovereignty of the nation dissolves in a borderless world of Empire. It can be countered, as I shall indicate below, that these two tendencies are not as distinct as Hardt and Negri suppose, and that both form key components of US imperialism. It can be argued that US imperialism has always followed a double movement of erecting and policing boundaries, and of breaking down the borders of others both internally and externally so as to open up new spaces for unfettered expansion.

I want to pursue their argument a little further on this question, since the theme is central to this chapter and has implications for subsequent chapters.

In Hardt and Negri's chapter on US sovereignty and network power, it is suggested that the US constitution is expressive of a democratically expansive tendency which is not exclusive. When it expands, this new sovereignty does not destroy the other powers it faces but opens itself to them. Perhaps, it is argued, the essential feature of this new imperial sovereignty is that '*its space is always open*' (Hardt & Negri 2000: 167; emphasis in original). But equally it is suggested that the utopia of open spaces hides a brutal form of subordination of the Amerindian population. The Native Americans were the 'negative foundation' of the US constitution since their exclusion and elimination were essential for its functioning and the facilitation of openness and colonizing expansion (2000: 170). Native Americans could be excluded because the republic did not need their labour, but with African Americans inclusion was necessary since black labour was an essential support to the new United States. African American slaves could be neither completely included nor entirely excluded, and this was reflected in the constitution itself, wherein the slave population counted in the determination of the number of representatives for each state in the House of Representatives, but at a ratio whereby one slave equalled three-fifths of a free person. For Hardt and Negri, black slavery was both an exception to and a foundation of the constitution, and such a contradiction posed a crisis for the new notion of US sovereignty because it blocked the free circulation and equality that animated its foundation.

Subsequently, Hardt and Negri suggest that in the Cold War period the United States was the author of direct and brutal imperialist projects, such as the repression of liberation struggles throughout the world. They then add that this imperialism was not really born in the 1960s, nor even with the beginning of the Cold War, but goes back to the Soviet revolution and perhaps earlier. 'Perhaps', they continue, 'what we have presented as *exceptions* to the development of imperial sovereignty should instead be linked together as a real tendency, an alternative within the history of the US Constitution'; or 'in other words, perhaps the root of these imperialist practices should be traced back to the very origins of the country, to black slavery and the genocidal wars against Native Americans' (Hardt & Negri 2000: 177). This surprising but revealing passage tends to undermine their perspective on the difference imperial sovereignty is supposed to entail.

Let us now consider the Jeffersonian notion of an 'Empire for Liberty' and offer an alternative approach to that of Hardt and Negri, one which makes no distinction between the imperial and the imperialist, and counters their idea of an open and inclusive Empire.

Roots of Empire

In 1801 Thomas Jefferson wrote as follows:

> However our present situation may restrain us within our own limits, it is impossible not to look forward to distant times, when our rapid multiplication will expand itself beyond those limits, and cover the whole northern, if not southern continent, with a people speaking the same language, governed in similar forms, and by similar laws; nor can we contemplate with satisfaction either blot or mixture on that surface. (qtd in Gardner et al. 1973: 64)

The desire for territorial expansion is conjoined with an imperialist sameness and order where any notion of 'mixture' or 'blot' is not to be gladly contemplated. There is little open space here for the recognition of difference. In the geopolitical formation of the United States, in the territorial constitution of its sovereignty, 'internal others' were subjected to a delineation that underpinned and legitimized an accelerating process of territorial expansion and subjugation.[1] The Empire for Liberty that Jefferson proposed was expansionist both internally and externally, and it carried within it a logic for the subordination of the indigenous other.

For example, in 1803 Jefferson, in a message to the Senate and House of Representatives, outlined a position on 'the Indian tribes residing within the limits of the United States' that effectively expressed such a logic. It was argued, for instance, that in order to provide an 'extension of territory which the rapid increase of our numbers will call for' two measures were needed. First, these indigenous peoples were to be encouraged to abandon hunting and devote themselves to the practices of stock raising, agriculture and domestic manufacture, so that the extensive forests they once saw as necessary for their livelihood would be exchanged for the means of improving their farms. Second, by multiplying trading houses among them, and by leading them to agriculture, manufactures and civilization, we would be preparing them ultimately to participate in the 'benefits of our Government' and therefore acting for 'their greatest good'.[2]

In his Second Inaugural Address of 1805, Jefferson went on to characterize resistance to the incursion of a superior civilization in terms of the presence within the aboriginal population of prejudices, ignorance, pride and the influence of 'interested and crafty individuals' who inculcated a 'sanctimonious reverence for the customs of their ancestors', and portrayed ignorance as safety and knowledge as replete with danger (see Richardson 1896: 378–82). These were the problems that faced the spread of Enlightenment and the exercise of reason.

In these and related passages, one can see the construction of significant governing representations which reflect a clear adherence to the idea of a civilizing mission coupled with the appropriation of territory. The establishment of a divide in the evaluation of different peoples and cultures is anchored to a belief in enlightened reason and the possibility of transforming the other. Through the diffusion of freedom, reason, science and moral advancement the natives would be transformed into a subordinated, dependent and inferior category of social subjects.

The Jeffersonian perspective provides a founding representation, traces of which can be found in the modernization theory of the 1950s, with its modern-traditional dichotomy and its prioritization of diffusion. Also, Jefferson's troublesome and 'crafty individuals' can be seen re-presented as the Communist subversives who were envisaged as attempting to destabilize and derail the rational process of the transition from the traditional to the modern society, while domestically those same individuals became, in the Cold War period, the 'enemy within', the organizers of 'un-American activities' (see chapter 3). But Jefferson also provided a rationale for nineteenth-century expansion that straddled both the immediate hinterland of the United States and also the societies to the South, pithily reflected in the possessive US definition of the Western hemisphere as a 'hemisphere to itself' (Niess 1990). As a way of pursuing this joint theme, it is helpful to outline some primary aspects of the importance of the frontier in the process of territorial expansion.

Territory and the Other – On the Place of the Frontier

At the beginning of the 1890s, the US Census Bureau declared that the frontier no longer existed. By the time of the massacre of Wounded Knee in 1890, Americans were settling the entire continent from the

Atlantic to the Pacific, and this conquest of nature had been driven by the explosive formation of an industrial economy. By 1890, for example, United States manufacturing production surpassed the combined total of England and Germany (Takaki 1993: 225), and by 1892 US foreign trade had already exceeded that of every country in the world except Great Britain. But the end of the frontier signified far more than economic expansion.

Frederick Jackson Turner (1962: 37–8), in his 1893 essay on the frontier in American history, argued that the process of settlement and colonization brought to life intellectual traits of profound importance – the emergence of a 'dominant individualism', a 'masterful grasp of material things', a 'practical, inventive turn of mind', a 'restless, nervous energy', which all reflected the specificity of the American intellect. Moreover, he did not limit his thesis to the internal territory of the United States, noting three years after his seminal paper that for nearly three centuries the dominant fact in American life has been expansion. For Turner, the demands for a 'vigorous foreign policy, for an interoceanic canal, for a revival of our power upon the seas, and for the extension of American influence to outlying islands and adjoining countries are indications that the movement will continue' (Turner 1896: 296).

In a similar vein, Theodore Roosevelt (1889: 26–7), in his four-volume examination of the frontier, entitled *The Winning of the West*, portrayed the frontier farmers and 'warlike borderers' as an 'oncoming white flood', while their adversaries, the native Indians, were considered to be the 'most formidable savage foes ever encountered by colonists of European stock' (ibid.: 17). Roosevelt's notion of an 'oncoming white flood' being associated with a civilizing mission found an international expression in what came to be known as the Roosevelt Corollary of 1904, which will be mentioned below.

By the beginning of the twentieth century, as the waves of colonization and settlement had passed their peak, the earlier Jeffersonian objective of separating the Indians from their land, of incorporating and assimilating the Indian into an advancing civilization, had taken its toll. War and subsequent treaties resulted in Native America being constricted to about 2.5 per cent of its original land base within 48 contiguous states of the union (Rickard 1998: 58), and the violent appropriation of land and subsequent confinement of native Indian communities to limited reservations gave another darker expression to the meaning of the frontier (Slotkin 1998; Takaki 1993: 228–45; and Zinn 1980: 124–46). The reality of war and violence was customarily represented as an unavoidable, preordained consequence of the beneficial march of a new civilization (see, for example, Tocqueville 1990: 25).

The expansion of the frontier, and the territorial constitution of the United States as we know it today, had another dimension which was also particularly relevant to the later projection of power and hegemony in the societies of the South, and especially in the context of US–Latin American relations.

On the eve of the US–Mexican War, and in the wake of the annexation of Texas from Mexico, a pivotal cause of the war, notions of 'Manifest Destiny' came to circulate in the worlds of journalism and politics.[3] John L. O'Sullivan, the editor of the *Democratic Review*, and the originator of the term, had already written in 1839 of a boundless future for America, asserting that 'in its magnificent domain of space and time, the nation of many nations is destined to manifest to mankind the excellence of divine principles'. It was six years later in 1845, in relation to continuing opposition to the annexation of Texas into the Union, that O'Sullivan wrote of 'our manifest destiny to overspread the continent allotted by Providence for the free development of our yearly multiplying millions' (both quotations in Pratt 1927: 797–8). The doctrine of 'manifest destiny' embraced a belief in American Anglo-Saxon superiority, and it was deployed to justify war and the appropriation of approximately 50 per cent of Mexico's original territory. Furthermore, as with accompanying characterizations of the native Indians' purported lack of efficient utilization of their natural resources, it was observed by President Polk, at the end of the War in 1848, that the territories ceded by Mexico had remained and would have continued to remain of 'little value to her or to any nation, while as part of our Union they will be productive of vast benefits to the United States, to the commercial world, and the general interests of mankind' (quoted in Gantenbein 1950: 560).

Territorial expansion into Mexico was characterized by a contentious debate over the perceived advantages and disadvantages of incorporating a people deemed to be so palpably inferior. For example, Secretary of State James Buchanan often expressed his fear of the admission of any large number of Mexicans to the Union, and in the wake of the 1848 Treaty of Guadalupe Hidalgo, one newspaper, the *Louisville Democrat*, expressed the opinion that the United States had obtained 'not the best boundary, but all the territory of value that we can get without taking the people' (quoted in Horsman 1981: 245–6). Efforts to acquire all of Mexico in the 1846–8 war raised strong racialist objections to having the Mexicans brought into the Union. Rhode Island Senator John Clarke, who epitomized the Northern Whig position, expressed the view that a 'degraded population, far inferior to the Atzec race', and with no 'fixed principles of government' should not be incorporated into the Union, since such an act would be 'fatally destructive to the institutions of our country' (Schoultz 1999: 35).

It was a widely-held belief throughout the nineteenth century, and particularly before the Civil War, that people who were deemed not to be capable of self-government should not participate in the governing of Anglo-Saxons. In this context, the later expansion which took place post-1898, and which concerned the acquisition of non-contiguous territory inhabited by races considered by the dominant sentiment within the United States to be inferior, was accompanied by an evasion of the US constitutional principle that all citizens of a republic ought to enjoy an equality of rights.[4] In other words, as Weston (1972) has shown in his analysis of the influence of racial assumptions on the foreign policy of the United States, the ideology of racial superiority tended to compromise the principles of equality of rights and opportunity, a factor that also impinged on domestic politics. Dominant attitudes towards countries with substantial African populations, such as Cuba and Haiti, were continually laced with racial assumptions of Anglo-Saxon superiority and the associated opposition towards any annexation of their territories (Hunt 1987). This did not mean that the civilizing mission towards these other peoples and societies was in any way dimmed, but rather that the prosecution of such a mission was achieved through a number of different means.

Governing Visions and the Geopolitics of Intervention

The historical context

One of the issues that often overshadows much of the debate on imperialism and post-colonial theory concerns the roots of expansionism, or in Blaut's (1993) terminology 'diffusionism'. In traditional Marxist accounts one would invariably encounter an emphasis on the economic – the search for new markets, raw materials, cheap labour or higher profits. Clearly in the case of the United States, by the end of the nineteenth century, as the internal frontier reached its point of closure, and as domestic overproduction, agrarian unrest and labour conflicts became more acute, the drive to expand and seek out new territories for economic development became more manifest (LaFeber 1963). But it would be somewhat limiting to interpret expansionism as only being driven by social and economic pressures. These pressures were certainly a factor, but before the 1880s there was ample evidence of emerging US interventionism that related to a wider set of issues than only the economic.

Expansionism was justified in relation to a civilizing mission. Josiah Strong, a key spokesman for the American Evangelical Alliance, in his vision of the 'missionary frontier' argued that the Anglo-Saxons with their 'genius for colonizing' would move down upon Mexico and Central and South America and out over Africa and beyond, bringing Western civilization and Christianity to a backward and heathen world. Similarly, for Alfred T. Mahan, the development of sea power and access to overseas markets in Asia and Latin America, via the establishment of strategic bases, were linked to the prowess of Western civilization. Mahan held the conviction that external threats could be overcome through the spread of progress, and other 'stagnant societies' could be regenerated through American expansion (Healy 1970: 129–30). In the later part of the nineteenth century, notions of the inevitable and beneficial spread of Western progress were also influenced by a fashionable social Darwinism, linked to a belief in America's geopolitical predestination (Weinberg 1963).

Such views were also very much part of the governing representations of the day. President McKinley, for example, at the end of the 1890s, in a speech given in Ohio, against the backdrop of the take-over of the Philippines, proclaimed that, 'whatever obligation shall justly come from this strife for humanity, we must take up and perform and as free, strong, brave people, accept the trust which civilization puts upon us' (May 1973: 259). Similarly, a few years earlier, in 1895, Secretary of State Richard Olney, in a letter to the British ambassador in the US concerning a border dispute between Venezuela and Britain, made a connection between civilization and US power, noting that 'the United States is practically sovereign on this continent, and its fiat is law upon the subjects to which it confines its interposition' (Holden & Zolov 2000: 66).

In the light of Olney's dictum, it can be noted that an expansion of spatial power or the establishment of a new spatial-political order needs a justification, an ensemble of ideas and concepts that can provide a moral and cultural foundation. Moreover, in the context of relations with other societies, and specifically in the Americas, remembering Jefferson's notion of the United States having a 'hemisphere to itself', the construction of a geopolitical identity included the positing of difference as inferiority *and* danger. The outside world contained threats to security and to the diffusion of mission. The perceived threat of disorder and chaos to the rule of the emerging American Empire could be taken as an example of the key relation between the perception of threat and the geopolitics of intervention to maintain a sense of security.

It can be argued here that any discussion of threats to order and stability must be linked to discourses of identity and difference. What exactly is being threatened? What are the discourses or regimes of truth that are immanent in the power relations that seek to preserve order? In the case of nineteenth-century US power, the spread of progress and a civilizational mission were predominantly envisaged as being rooted in a specifically 'American destiny'. There were the pressures of economic expansion, but the USA's representation of itself to the world, the construction of a project of leadership (backed by the capacity and willingness to deploy force) was crucial to any understanding of the geopolitics of interventionism. Threats and perceptions of disorder are predicated on governing visions which are one expression of the complex intersections of power and cultural representation. But these visions are also a reflection of a hegemonic ambition.

Further, in a context that is international, where the intersubjectivity that is a pivotal part of power relations stretches across national boundaries, and therefore national cultures, and where the attempt to develop a hegemonic project comes up against nationalist opposition, one kind of counter-geopolitics, the resistance to imperial persuasion, has been strikingly resilient, even if it has never been the only tendency, as will be mentioned below. What needs to be remembered, as I shall suggest below, is that in any account of the power/discourse intersection, the effectiveness of counter-discourses or counter-representations to Empire ought to be included as a significant part of the analysis.

With these points in mind, I now want to pursue the theme of governing visions and the geopolitics of interventionism by taking the example of the aftermath of the Spanish-American War of 1898, and its implications for the way we might think about the emerging global power of the United States at the dawn of the twentieth century.

US power in 1898 and beyond: the varying cases of the Philippines, Puerto Rico and Cuba

At the end of the 1890s, the United States annexed the nominally independent Hawaiian Islands, and in 1900 Hawaii was made a full-fledged territory with American citizenship conferred on all citizens of the islands. In 1898, following the Spanish-American War, Spain and the United States signed the Treaty of Paris, through which Spain ceded Cuba, Puerto Rico, Guam and the Philippine archipelago to the US. In the cases of Hawaii and the Philippines, annexation had been strongly influenced by the perceived danger of another Pacific power, Japan, securing control of these island territories. With specific reference to the Philippines, the islands could become a key military base from which to protect US interests in Asia. In the contrasting case of Cuba, with its proximity to the United States and the absence of rival geopolitical powers, there was a deeply rooted belief that the island belonged naturally within the orbit of US power.[5] Moreover, in contrast to the Philippines, Cuba, as a result of a protracted independence struggle, had run up large debts which under annexation would have become a direct US responsibility. In this overall context, anti-annexation tendencies within the United States argued that the deployment of a colonizing power would violate the founding ethos of the Republic. In fact, in the US Senate, after a long debate over the adoption of a policy of 'imperialism', exemplified in the annexation of the Philippines, the consent to annexation was only ratified by a close vote in early 1899. After the defeat of the Filipino movement for independence in 1902, a civil government was established in 1901 under the Philippine Commission, which operated as a legislature under the US Governor General.

Although territorial expansion had been an intrinsic part of United States development before 1898, the Spanish-American War brought in its wake a qualitatively different form of expansion that entailed the acquisition or control of sizeable dependencies not contiguous to the homeland, as had been the case with Mexico, and inhabited by other peoples with their own histories, cultures and political institutions. US leaders had to determine how these new territories should be governed: whether, for instance, they should be colonies in an American Empire, or rather more loosely-tied dependencies, with the possibility of some degree of self-government under US tutelage. In an important sense, the answers to these questions were gradually worked out in the cases of Cuba, the Philippines and Puerto Rico, where different approaches were applied. These three examples, because of differences in their geopolitical position, histories and relations to US power, were dealt with in comparable but specific ways, whereby different mechanisms of geopolitical power were applied by the United States.

In the case of the Philippines, formal annexation and colonial government went together with a Filipino insurrection. Rebel leader Aguinaldo's independent republic was backed by 10,000 well-armed insurgents who did not accept President McKinley's promise that they were to be 'liberated'. American troops, more than 120,000 overall, and the Filipino insurgents waged war for nearly 3 years, and when in 1902 the US War Department declared the insurgency at an end, an estimated 20,000 Filipino insurgents had been killed in battle, and as many as 200,000 civilians were dead of hunger or disease. In the words of Karney (1989: 12), 'it was an unalloyed American conquest of territory and among the cruellest conflicts in the annals of Western imperialism'. The material and moral costs of the war and the internal divisions that it provoked in the United States acted as a powerful brake on the development of any future strategy of territorial annexation.

At the same time, it needs to be remembered that opposition to annexation and imperialist power was not necessarily anchored in any anti-racist conviction (see Lasch 1973).[6] Industrialists like Andrew Carnegie were of the opinion that costly foreign ventures would divert the United States from the development of the domestic economy, and northern factory workers and southern farmers who supported the Democratic Party tended to be isolationist. In addition, and significantly, annexation was frequently opposed on the grounds that it entailed the incorporation of inferior races into the American Republic, as had been argued in the Mexican case.

In the case of Puerto Rico, also ceded to the United States after the Spanish-American War, the Foraker Act of 1900 authorized the President to appoint a governor and executive council. The council, which consisted of the 6 heads of administrative departments plus 5 Puerto Ricans, acted as the upper house of the legislature, while the lower was elected by the people. The inhabitants were not given US citizenship until 1917, when following the Jones Act, Puerto Ricans were empowered to elect a Senate; but the governor, appointed by Washington, still had veto power, and the island's sovereignty still depended upon the authority of the US Congress. In what were referred to as 'insular cases', the US Supreme Court upheld the right of Congress to treat newly acquired territories as not fully incorporated into the United States and hence not subject to all of the provisions of the Constitution. Puerto Rico was ambiguously classified as an 'unincorporated territory'.

As Kaplan (2002) has effectively argued, in the Puerto Rican case, the pursuit of imperial desire carried the risk of absorbing aliens into the domestic sphere, and therefore it was deemed necessary to assign Puerto Rico a distinct legal status to foreclose this possibility. So it was that Puerto Rico went from being an 'unincorporated territory' to being a 'self-governing Commonwealth' from 1952. The initial period of US occupation was marked by conflict, violence and the loss of basic civil rights. For the Puerto Rican sociologist González-Cruz (1998), five measures stood out in this process: militarization, control over the means of communication, defence of the privileged classes, repression of political movements critical of or opposed to the regime, and the establishment of new local government structures. These features were also to be found in the Cuban case, as I shall mention below.

The Spanish-American War of 1898 was sparked by conflict in Cuba. The Cubans had been fighting for independence from Spain and in the United States anti-Spanish feeling was fuelled by Cuban groups and US newspapers such as Hearst's *Journal* and Pulitzer's *World*. Reacting to popular feeling, and following the sinking of the US warship the *Maine* in Havana harbour in 1898, an event which was erroneously blamed on Spanish agents, Congress authorized war in April 1898 and Spanish capitulation ended a brief struggle. Formally placed under United States military government in 1899, Cuba became a republic under US tutelage in 1902. The Teller Amendment of 1898 prevented the United States from openly annexing the island. Passed by Congress when US public opinion regarded military assistance for Cuba's struggle against Spain as heroic, the amendment stated that the United States disclaimed any intention of exercising sovereignty or control over the island except for pacification thereof, and asserted its determination, when the pacification was accomplished, 'to leave the government and control of the island to its people' (quoted in Pérez 1997: 96). The Teller Amendment eased some consciences, and was supported by those politicians who did not want the United States to annex an island inhabited by an 'alien' people (see Smith 2000: 49), but also the amendment aimed to protect US sugar producers from imports of cheap Cuban sugar, Senator Teller himself coming from the sugar-beet state of Colorado.

As Cuba was brought into the orbit of US power, the notion of 'manifest destiny' was invoked, with one writer asserting that 'it is manifest destiny that the commerce and the progress of the island shall follow American channels and adopt American forms' (Benjamin 1990: 54). As a result of US intervention and occupation, Cuba ceded territory for the establishment of a US naval base at Guantánamo Bay, agreed to a significant curtailment of its national sovereignty and authorized future US interventions. The Platt Amendment which was enacted into law by the US Congress in 1901, and reluctantly accepted by the new Cuban government in 1902, became a key part of the 1903 Permanent Treaty between the two countries. Through this Treaty key restrictions were imposed on Cuba's conduct of foreign relations. Article 3 was particularly crucial, stating that 'the government of Cuba consents that the United States may exercise the right to intervene for the preservation of Cuban independence.' This article was closely connected to article 7, which stated that 'to enable the United States to maintain the independence of Cuba and to protect the people thereof, as well as for its own defense, the government of Cuba will sell or lease to the United States lands necessary for coaling or naval stations' (Holden & Zolov 2000: 81–2). The Platt Amendment was a source of resentment inside Cuba for a considerable period of time and was not abrogated until 1934 under President Roosevelt; even then the substance of article 7 remained in force with Guantánamo Bay remaining under US control.

In the Cuban case, the United States developed a strategy that provided an important backdrop for the future of geopolitical interventions in Central America and the Caribbean. The strategy rested on five interrelated objectives.

The first of these, which was basic to the others, consisted in the establishment of informal protectorates which provided a political space for internal self-government. When it was deemed necessary, as in cases of internal revolts and acute political instability, military occupation by United States forces became an option, but never the assumption of formal sovereignty. In Cuba, for example, in the early part of the twentieth century, the United States intervened militarily in 1906–9, 1912 and 1917–22 to restore order and protect 'American interests', while controlling revolutionary activity (Williams 1980: 138–41).

Second, through trade, treaties or financial arrangements, a series of strong economic ties was woven between the United States and the dependent country. These ties were also and importantly put into place through land purchases by North Americans, so that by 1905 13,000 US citizens had bought land in Cuba worth US$50 million, much of the land being used for sugar production, and by 1913 US investments in Cuba would total about $220 million – 18 per cent of US investments in the whole of Latin America (Thomas 2001: 365).

Third, by means of investment and a variety of projects of improvement – the diffusion of progress – new forms of economic and social involvement were put into place, including improvements in health care, the reform of public education and a modernizing programme of public works. This included in the Cuban example the transfer of over 1,000 Cuban teachers to Harvard for training in US teaching methods, together with the establishment by Protestant evangelists of almost 90 schools between 1898 and 1901.

Fourth, there was a reterritorialization of administrative power, including the introduction of an American version of local self-government, and the central problem here, as one commentator noted for the Cuban case, lay 'in the attempt to engraft the Anglo-Saxon principle of local self-government on an Iberian system to which it was wholly foreign' (quoted in Healy 1963: 184).

Finally, through the disbandment or re-organization of particular national institutions, such as those that had been developed by the Cuban independence movement – the Liberation Army, the provisional government and the Cuban Revolutionary Party, originally founded by José Martí – and the initiation of processes of cultural penetration and subordination, attempts were made to Americanize the 'subject peoples'.

This attempted 'Americanization' of subordinated peoples went to-gether with a protracted series of interventions, especially evident in Central America and the Caribbean. Between 1898 and 1934 the United States launched more than 30 military interventions in Latin America, and the vast majority of these interventions took place in the Caribbean basin. Most US interventions displayed a consistent pattern. Military forces would arrive and depose indigenous rulers, often with a minimum of force, install a hand-picked provisional government, supervise national elections, and then depart, mission accomplished (Smith 2000: 51–2). Apart from the three cases of the Philippines (a colony), Puerto Rico (an 'unincorporated territory') and Cuba (a semi-protectorate until 1934), which were specific in the sense of being acquired in the wake of the Spanish-American War, and which exhibited different mechanisms of US geopolitical power, interventions in countries to the south of the United States clearly related to Roosevelt's invocation of an 'inter-national police power', popularly known as the wielding of a 'big stick'. The Roosevelt Corollary of 1904, as it became known, provided an instructive example of the close association between governing visions and the geopolitics of intervention. Roosevelt's invocation of an 'inter-national police power' was linked to his negative view of the 'weak and chaotic people south of us', who would, if need be, have to be policed in the interests of 'order and civilization' (quoted in Niess 1990: 76). The

Roosevelt Corollary was itself linked to the Monroe Doctrine of 1823, which provided the initial codification of emerging US power and the first phase of containment, limited to the Americas (Slater 1999b). But there were other governing visions which emerged in the period before the Second World War which were also highly significant.

Dollar Diplomacy and Democratic Mission

President W. H. Taft, who followed Roosevelt to the White House in 1909, and who was also a Republican, was credited with the new notion of 'dollar diplomacy'. In contrast to Theodore Roosevelt, who approached the question of the Panama Canal more in terms of the flow of imperial traffic, Taft emphasized the fact that it was the United States that owned the canal, having built it with 'our money'. Foreign policy for Taft needed to be connected to changes in political economy. Since the United States had now 'accumulated a surplus of capital beyond the requirements of internal development', it needed to channel its energies abroad, responding to the modern ideas of commercial intercourse, and substituting 'dollars for bullets' (see Williams 1980: 132–3).

This stress on the importance of extending US trade and investment abroad was coupled with a policy of US representatives taking charge of customs houses in countries such as the Dominican Republic, Nicaragua and Haiti, so that payment on rescheduled debts could be more effectively guaranteed. This economic policy went hand in hand with military occupation. US financial control in the Dominican Republic, for example, led to close political supervision, and an uprising in 1911–12 led Washington to propose the resignation of a provisional president. In the case of Nicaragua in 1911 a bilateral agreement specified terms for the authorization of US$15 million in loans, and Nicaragua pledged not to alter customs duties without US approval. In the example of Haiti, between 1908 and 1915 there were 7 presidents and about 20 uprisings, and in 1915 the US marines invaded. In the same year, Washington imposed treaty terms on the new government, including control of the customs house and the US appointment of a financial adviser and US supervision of public works. For an expanding system of investments in export commodities, the United States required order and security, and 'dollar diplomacy' went together with military intervention to secure such stability.

In 1913 Woodrow Wilson assumed the presidency, and in foreign policy more attention came to be given to democracy and self-government, the significance of the latter term dating back to the founding fathers and especially to Thomas Jefferson. While I have underscored the difference between Roosevelt and Taft, a difference in which the former has been associated with the importance of military power and the latter with the capitalist imperative of economic expansion, one ought not to assume that Roosevelt did not appreciate the place of economic capacity, nor that Taft did not realize the necessity of military prowess. In the case of Wilson, his perspective sought to bring these two sources of power more closely together and to do so by also re-emphasizing the 'American mission' – his 1917 war aim being to 'make the world safe for democracy'. It was President Wilson who called for and began to build a navy second to none, effectively supported the expansion of US economic strength, together with a strong advocacy of free trade,[7] and intervened repeatedly both militarily and politically in societies considered to be backward and disorderly (e.g. Mexico). In fact, Wilson, the son of a Presbyterian minister, had already stated in the early years of the century that it was the 'peculiar duty' of the United States to teach the peoples who inhabited its new colonial frontiers the need for 'order and self-control' (Black 1988: 280).

In the context of Mexico, Central America and the Caribbean, the Wilsonian period was characterized by continual interventions – in Cuba from 1917 to 1922, in the Dominican Republic in 1914 and again in 1916–24, in Guatemala in 1920, in Haiti from 1915 lasting through until 1934, in Honduras in 1919, and in Mexico in 1914, 1916–17 and 1918–19. Wilson believed in preserving order and confronting any threats to US interests abroad, but also he was an advocate of the notion that 'self-government' was possible for all peoples, even those less civilized, provided they were correctly instructed – he asserted, for instance, that 'when properly directed there is no people not fitted for self-government' (Weber 1995: 84). For Wilson, the United States was special because it had seen visions that other nations had not seen, and its mission was to be the 'light of the world' and to lead the world in the defence of the rights of peoples and the rights of free nations.

At home, however, during the period of the First World War, Wilson's universalism tended to be accompanied by an illiberal approach to opposition within the United States once the country had entered the war. Criticism of the war became illegal, and numerous people were imprisoned, while perceived dissenters from the American way of life and 'hyphenated' Americans were subjected to discrimination (Stephanson 1995: 119). Wilson saw no contradiction in his universalist conception of democracy. However, while he expressed public criticism of imperialism, noting in his fourteenth point for world peace and democracy, written in 1918, that political independence and territorial integrity must be extended to great and small states alike (Maidment & Dawson 1994: 261), in practice he sanctioned the sending of armies into Mexico and revolutionary Russia, as well as presiding over the military interventions in Latin America that I have mentioned above.

How do we then explain the emphasis on a mission for democracy and self-government? Clearly this was part of a project aimed at developing a geopolitical hegemony for the United States. The emphasis on self-government and a democratic future written according to an American imprint, represented an early twentieth-century attempt to 'globalize the Monroe Doctrine'. Furthermore, such an emphasis reflected the attempt to transcend the contradiction between support for the post-coloniality of self-determination and the reality of the United States as an imperial power. However, such a projected hegemony did not go unchallenged.

The Vitality of Counter-representations

In the case of Cuba, the imposition of the Platt Amendment and the extension of US power and influence in the island fuelled the fires of opposition and helped nurture the rise of nationalism and anti-imperialist sentiment (Foner 1972: 593–612). Resistances to US power emerged out of the rich vein of the independence struggle against Spanish colonialism, which provided the foundation for new kinds of counter-representations of North–South relations in the Americas. In particular, José Martí, writer and revolutionary nationalist, developed a view of US–Latin American relations that continually stressed the need for autonomy when faced by the 'colossus of the North'.

Writing in 1891, four years before his death on the battlefield, Martí argued that although Spanish America had almost entirely freed itself from the first metropolis, a new and much more powerful metropolis was overtaking it under the guise of economic penetration, and by political and where necessary military means. In the context of economic penetration, Martí (1961: 19–29), in a stance not without contemporary relevance, was opposed to proposals for a trade agreement between Mexico and the United States, whose only beneficiary, in his view, would be the United States. Under the influence of Simón Bolívar, Martí believed that it was crucial that Spanish America, and specifically his own Cuba, should not commit the error of regarding Europe or North America as the foreign model to which it should adjust – ideas, for Martí, must come from deep roots and compatible soil if they are to develop a firm footing and prosper, a view that presciently anticipated the 1960s Latin American literature of *dependencia*, which I shall examine in chapter 5.

Also concerned by the impact of imperialist politics on Latin America, the Argentinian intellectual José Ingenieros, when visiting Mexico in 1922, called for the formation of a Latin American Union to forge a confederation of Latin nations allied in self-defence against the United States. For Ingenieros, the United States was not only a powerful neighbour but also a 'meddlesome friend' and a danger to Latin America. This danger was not just related to military interventions but to the mortgaging of national independence through loans. He concluded his speech in Mexico City by stating that the key issue for Latin Americans was national defence – 'American capitalism,' he wrote, 'seeks to capture the sources of our wealth, with the right to intervene in order to protect its investments' (see Holden & Zolov 2000: 124–5).

Apart from the example of Latin American writers, alternative representations of international relations also emerged in the world of diplomacy. For instance, in 1896 the Argentinian diplomat Carlos Calvo challenged the major European powers of the day who were claiming an international legal right to protect their nationals and their nationals' property anywhere in the world, and if necessary by deploying armed force. Calvo stated that in his view, and according to strict international law, the recovery of debts and the pursuit of private claims did not justify the armed intervention of governments, since the territories of sovereign states should be recognized as being universally inviolable, whether those states were European or Latin American. Calvo's position was followed a few years later by the Drago Doctrine of 1902, which similarly stated, *inter alia*, that public international law was based on the key assumption that all states are entities in law, perfectly equal to one another and therefore entitled to the same consideration and respect.

Both these initiatives in the field of international law were primarily addressed to the European powers, but at the same time, and in relation to the position of the United States, they expressed an independent Latin American stance. At the Third Pan-American Conference in 1906, held in Rio de Janeiro, Washington rejected the Drago Doctrine, just as it had previously rejected the Calvo proposal. Moreover, in the 1920s, President Coolidge was of the view that the person and property of a citizen are part of the general domain of the nation, i.e. the US nation, even when abroad. This thesis formed the basis of the Evart Doctrine, which claimed legal immunity for US citizens and their business activities in Latin America, in violation of the national sovereignty of the affected countries (Niess 1990: 78). This overarching position of the US government provided another modality of justification for interventionism in Latin America, and was a continuing source of friction in US–Latin American relations.

Finally, it needs to be emphasized that the counter-representations of Latin American intellectuals and diplomats that surfaced from the late nineteenth century onward were a key component of the development of Latin American nationalisms and the insurgency of the times. The Mexican Revolution (1910–20) was clearly the main manifestation of such insurgency, with a peasant-led movement of nationalist politics bringing in its wake profound changes in the internal structure of power relations with far-reaching implications for the architecture of US–Mexico relations from the 1920s onwards. Also, in the wake of the Russian Revolution of 1917, the 1920s saw the establishment of Communist Parties in many parts of the Latin South, and the development of Marxist thought. In addition, US interventions and occupations did not go unchallenged, as was clearly seen in the case of the Dominican Republic, where between 1916 and 1924, in the eastern part of the island, US marines faced a peasant-based guerrilla insurgency (Calder 1978). Moreover, in the economic domain, the 1920s and 1930s witnessed a wave of demonstrations against US corporations. Not only in Central America and the Caribbean but also in the economies of South America, opposition to US capital became more overt. In 1925 in Chile, for instance, the vast majority of the 7,000 mine workers at the US-owned Chuquicamata copper plant went on strike in protest against the company's prohibition of workers' meetings and circulation of labour publications, and the workers denounced the 'imperialist domination' of the United States. In another example, in 1930 Peru, workers, aided by urban radicals and elements from the military, paralysed the US-owned Cerro de Pasco Mining Corporation for four months (O'Brien 1996).

The emergence of new currents of nationalism and resistance to US hegemony affected the direction of US foreign policy, and in the 1930s President Franklin D. Roosevelt initiated his 'Good Neighbour Policy', which was an important response to the new geopolitical climate.

Mutations in the Quest for US Hegemony

Even though Roosevelt, as assistant secretary of the navy in the Woodrow Wilson administration, had played a key role in the US occupations of Haiti, the Dominican Republic and the Mexican port of Veracruz, in the 1930s he denounced the policy of intervention. In his inaugural speech of March 1933 he declared that his world policy would be that of the 'good neighbour'. In his address to the Pan-American Union in Washington he asserted that the essential qualities of a true Pan-Americanism must be the same as those of a good neighbour, namely mutual understanding and a sympathetic appreciation of the other's point of view. He then went on to state that the Monroe Doctrine ought to be directed at the maintenance of independence by the peoples of the continent. Together with the Pan-American doctrine of continental self-defence, the peoples of the American Republics ought to understand that each Republic must recognize the independence of every other Republic. Roosevelt declared that 'each one of us must grow by an advancement of civilization and social well-being and not by the acquisition of territory at the expense of any neighbour'. The address also included a plea for free trade, noting that the healthy flow of trade between the peoples of the American republics should not be obstructed by 'unnecessary and artificial barriers'.[8]

Roosevelt had come to the view, and in Washington he was not alone in this, that overt military intervention in Latin America was a source of political discontent and turbulence, which in its turn was disadvantageous to US business and a stable macro-economic climate. This was also linked to the increasing importance of trade and investment between the United States and Latin America, and in particular to the growth in US foreign direct investments in Latin America, rising from an estimated US$300 million in 1897 to $3.52 billion in 1929 (Niess 1990: 206 and 121).

Roosevelt's 'good neighbour' approach to Latin America – which, as will be seen in subsequent chapters, found echoes in the Alliance for Progress of 1961 and President Clinton's view of the North American Free Trade Agreement (NAFTA) of 1994 – gave priority to mutually beneficial relations between the northern and southern parts of the Americas in a context of free trade and political cooperation. Not only were the economic ties that bind being drawn closer between the north and south of the Americas, as reflected in trade and investment patterns, but also these ties were another part of the US 'mission' to export its way of life (O'Brien 1996: 251).

The 'good neighbour' policy was a significant attempt to provide a sense of political, economic, moral and cultural leadership for the Americas. It aimed at outlining the guidance and direction that was deemed necessary for Inter-American relations in turbulent times. It highlighted the importance of persuasion, of mutual understanding, responsibilities and respect, of a joint project of progress and friendship across the Americas, through which all the peoples of the 'American Family of Nations' would benefit and prosper. It underscored the key role of free trade, a point which Roosevelt repeated in a speech to the Inter-American Conference in Buenos Aires in 1936. Free trade and commerce were also intimately linked to a Wilsonian stress on democracy. Further, in the 'Good Neighbour' address as well as in other speeches given in the 1930s, Roosevelt returned to the importance of the 'spiritual solidarity' and 'spiritual unity' of the Americas, which he saw as an integral part of faith in God, and faith and spirit were invoked as being crucial to the 'Western Hemisphere' and the 'Western World'.

Statements on trade and commerce, democracy, religion and a spirit of mutual obligation, respect and understanding formed the key components of the Rooseveltian perspective. His vision was also associated with a new style, the US President being presented as a 'friendly uncle' figure offering a 'new deal' for the Latin American nations. In general, Roosevelt's intervention was an attempt to build a hegemonic discourse for the Americas, a way of bringing together the 'family of American nations' into a US-led project which expressed a multi-dimensional power. It was an attempt to nurture Latin American consent for the leadership role of the United States, and at the same time it was also a response to the development of Latin American nationalism and anti-imperialist sentiment. In practice, however, although 'gunboat diplomacy' faded from view, and concrete steps were taken to end the era of protectorates (as exemplified in the Cuban case with the abrogation of the Platt Amendment in 1934), dictatorships in countries such as the Dominican Republic, El Salvador, Guatemala, Honduras and Nicaragua were still supported (Black 1988). As a consequence and by the end of the 1930s, with US support for the strengthening of the armed forces in Latin America, Latin American opinion became much more critical of the 'good neighbour' notion, so that at the end of the 1930s a populist Peruvian leader, Haya de la Torre, could refer to the Roosevelt Administration as 'the good neighbor of tyrants' (Rosenberg 1982: 227).

Imperial Power and the Mission of 'America'

At the beginning of the chapter, I outlined some features of Empire in relation to the emerging power of the United States, and some critical comments were offered on Hardt and Negri's vision of Empire as related to the US. One of the constituent characteristics of the emerging imperial power of the United States concerned the underlying presence of a sense of mission and destiny. From 'Manifest Destiny' and the Roosevelt Corollary, to 'dollar diplomacy', the Wilsonian view of spreading US-style democracy, and finally to the 'good neighbour' policy, there has been, albeit in varied degrees, a deep sense of moral obligation to spread US values and practices to other ostensibly less-civilized societies. This belief in mission, with its religious connection, found a culminating expression in Henry Luce's notion of the twentieth century being 'our century', 'The American Century'.

Luce (1941) defined America's place on the world stage in terms of it being the dynamic leader of world trade, the key guarantor of the freedom of the seas, the disseminator of scientific leadership and the 'Good Samaritan' for the entire world. For Luce, America was the haven for a love of freedom; and, above all, Americans were the inheritors of all the great principles of Western civilization – justice, the love of truth and the ideal of charity. From this foundation, Luce asserted that it was now time for the United States to be the powerhouse from which these great ideals could be spread throughout the world. Luce captured the mood of the time, and his ideas of America's destiny to spread its power beyond its own frontiers had already found expression in the history of US–Latin American relations. His notion of an 'American Century' was prescient and came to be further reflected in ideas of modernization in the post-Second World War period, as we shall see in the following chapter. How finally, then, might we interpret imperial power?

From the varying sections of this chapter it is possible to extract my own perspective on imperial power, seen in the context of US relations with the societies of the South, and especially with the peoples of Latin America. It is possible to specify three interwoven elements of imperial power, which not only characterize the pre-1940 period, but also the subsequent post-Second World War era. They can be viewed as one approach to the formative elements of US imperial power.[9]

First, there is an enduring *invasiveness* or desire and capacity to penetrate other societies and cultures, where the penetration is multi-dimensional and affects the psychological, political, economic, financial and cultural realms in an intersecting manner. This element is part and parcel of the will to *impose* on the other a set of values, imaginations and practices, which are deemed to be superior – to ostensibly save the other from itself, to impose an external identity onto a recipient society or culture which is represented as in need of progress, or order, or civilization, or improvement, or reform or democracy, thus denying the other society its right to make its own destiny. In the context of the earlier section on governing visions and interventions, it was shown how these imposing representations are crucially imbricated with the legitimization of interventions, even though the mechanisms of these interventions will vary from one society to another, as was briefly illustrated in the cases of the Philippines, Puerto Rico and Cuba. This second element is then inseparably entwined with the third, which is constituted by the *lack of respect* for the other, being manifested in subordinating modes of representation, of negative essentializations which erase or belittle the complexities, differences, heterogeneities and *intrinsic value* of the other culture or society.

These three elements also need to be seen in a context of the vitality of the counter-representations emerging out of Latin America, and an important dimension of the post-colonial perspective developed in this chapter has been to underline the significance of these counter-visions, a theme that will be continued in the next chapter but more specifically in chapter 5, on dependency perspectives. As for the three elements of imperial power, the next chapter on modernization theory in the Cold War era will show that these elements have been accompanied by a limiting portrayal of the imperial heartland, an uncritical reading of the relations between imperial and imperialized societies, and a tendency to occlude the differential histories of the subordinated societies.

Part II
Waves of Western Theory

3

Modernizing the Other and the Three Worlds of Development

'Where the concepts and traditions of popular government are too weak to absorb successfully the intensity of the communist attack, then we must concede that harsh governmental measures of repression may be the only answer...'
                                    – G. F. Kennan, 1950 (in Holden & Zolov 2000: 196)

'Western civilization has made an immense contribution to the welfare of the world...the Western nations can feel that their greatest success was to have brought to much of the world a knowledge, a political freedom, and an economic opportunity which it had never before enjoyed.'
                                    – John Foster Dulles, 1956 (in US Department of State 1971: 31)

Introduction

As we have seen in the previous chapter, the emergence of the United States as a global power went together with a projection of notions of civilization, progress, democracy and order that posited a subordinate place for the societies of the non-West. The powers of expansion and intervention, both internally in the territorial constitution of the United States itself, and in a broader transnational mission of Empire, were intimately rooted in a vision of the United States as a driving force of Western civilization, diffusing its values to the presumed benefit of other non-Western societies (Cumings 1999). However, while US-modelled notions of civilization, progress, democracy and order continued to be transmitted in the period after the Second World War, and remained part of the Americanizing mission, other concepts came to receive greater emphasis.

From the 1950s onward notions of 'modernization' and 'development' came to be more closely associated with the portrayal of West/non-West encounters, whereas representations of civilization and order, although still present, as noted above in the Dulles quotation, became less prominent – they were no longer the master signifiers they had been before 1940. At the same time, democracy and order were resituated in a discursive context organized around the new signifiers of modernization and development. This does not mean, of course, that these terms had never been deployed before the Second World War, but rather that their visibility and discursive weight came to assume greater predominance in the post-War period. The post-War origins of the 'discourse of development' have been dealt with in considerable detail by Escobar (1995), while Patterson (1997) has traced the links between notions of Western civilization and modernization. Also, recent contributions (for example, Baber 2001 and Blaney & Inayatullah 2002) have revisited modernization theory in relation to Cold War politics and the conceptualization of international relations. What therefore still needs to be examined; or more precisely, what constitutes my own perspective?

First, in analysing the continuing intersections between geopolitical power and the cultural representations of other, non-Western societies, and particularly Latin America in the example of this study, it is important to keep in mind that the notion of modernization – or more specifically modernization theory – came to be closely associated with the nature and direction of US interventions in the Third World in the 1950s and 1960s. There was a specificity about this intersection which contrasts with earlier and later periods and needs to be understood in its geopolitical and historical context. It not only provides another important example of the interwoven nature of power, politics and representation but also illustrates the changing dynamic of US power as it impacted on the Third World.

Second, from a post-colonial perspective, and in the specific setting of this chapter, there are two analytical elements that can be usefully signalled:

a) The power to intervene was certainly not unaffected by the societies in which that invasive power was projected, since, as was noted in the previous chapter, resistances and oppositions to US hegemony altered the subsequent modalities of intervention, and this was particularly the case with respect to both the Cuban Revolution of 1959 and the

Vietnam War, set as they were in a broader context of accelerating geopolitical turbulence.

b) The geopolitics of intervention situated as it was in a Cold War context had an inside and an outside, since the Cold War had its chilling effect on domestic politics in the United States itself, and the phenomenon of 'containment culture' was a reflection of the inter-weaving of international and national concerns.

Third, modernization as an idea, and its association in the 1950s and 1960s with Americanization, was not new (Ceasar 1997: 168), and nor was it to disappear after the 1970s. As will be further shown in the next chapter, there are connections between neo-liberalism and moderniza-tion theory, as well as significant and often neglected differences. Furthermore, the term 'modernization' is frequently invoked today as if it had no history, and so in my own discussion an important objective is to highlight the historical specificity of the modernization idea in the Cold War era as part of a counter-geopolitics of memory.

In this chapter, I shall argue that the Occidental, and predominantly US enframing and deployment of modernization theory for the 'develop-ing world' was a reflection of a will to spatial power. It provided a legitimation for a whole series of incursions and penetrations that sought to subordinate, contain and assimilate the Third World as other. In the process it also put into place a vision of the West, and especially of the United States, which in some important aspects was re-asserted in later neo-liberal delineations of modern development, as well as in subsequent writings on globalization.

Developed for Diffusion: Situating the Expansionist Will to be Modern

Modernization was broadly defined as a universal process of change towards those types of social, economic and political systems that had developed in Western Europe and North America from the seventeenth to nineteenth centuries (Eisenstadt 1966 and Levy 1966). In the eco-nomic sphere, modernization was associated with the development of a high level of technology fostered by the systematic application of knowledge, the pursuit of which became the province of specialized scientific institutions. Culturally, a modern society was characterized by a growing differentiation of its value systems – religion, philosophy and science – and by the spread of literacy, secular education and a more complex intellectual division of labour. In terms of social change, modernization was connected to ideas of mobilization and differentiation; social mobilization, for example, was defined as a process in which major clusters of old social, economic and psychological commitments were eroded and broken as people became available for new patterns of socialization and behaviour (Deutsch 1963). As regards the political dimension, modernization theory tended to be constituted as the internationalization of the dominant paradigm employed in research on the American political system – the pragmatic-pluralist conception of the political process that defined liberal democracy in terms of high levels of participation, fair distributive outcomes and a consensual moral order (Apter 1987: 54–88).

Overall, there tended to be an association of modernization with societal transformation, so that for one American sociologist modernization was conceived as a total transformation of traditional or pre-modern societies into the types of technology and associated social organization that characterized the politically stable nations of the Western world (Moore 1966). Moreover, a distinction was sometimes drawn between notions of 'change' and 'development' that were envisaged as being incremental, and modernization which was posited as being tantamount to a systemic transformation (Halpern 1965). For one influential political scientist, modernization was a special kind of hope, and 'embodied within it are all the past revolutions of history and all the supreme human desires ... the modernization revolution is epic in its scale and moral in its significance' (Apter 1987: 54). There was also a belief, certainly expressed by Talcott Parsons, looking back at the 1950s and 1960s, that the trend toward modernization had become worldwide, and that in particular the elites of most non-modern societies accepted the crucial aspects of the values of modernity, especially economic development, education, political independence and some form of 'democracy' (Parsons 1971: 137).

In general, and notwithstanding the varying strands within it,[1] modernization theory as it was developed and deployed in the 1950s and 1960s can be characterized by the following features:

a) a linear view of history, in which Western countries were situated as being further along the path of modern development than Third World countries;

b) the positing of a crucial historical distinction between modern and traditional social systems, whereby the modern was defined in relation to a series of ostensibly primary attributes of Western societies – the scientific, secular, rational, innovative, democratic, open, plural, urban-industrial, achievement-oriented, and universally relevant – to be distinguished from the traditional, which was defined in relation to characteristics such as the particular, the religious, backwardness, the predominance of the rural, undeveloped divisions of labour, pre-democratic institutions, over-population, and the lack of capital, technology, entrepreneurship and modern values;

c) progress and development for the 'traditional society' would come about through the diffusion of modernization from the West to the non-West; and

d) successful transformations were not inevitable, and the passage from the traditional to the modern required a series of appropriate interventions – economic, financial, social, cultural, political and psychological.[2]

Modernization theory provided post-war society in the West, and especially the US, with a temporal and spatial identity, an identity that could only be effectively constructed in a relation of difference with another time and another space. In this sense the will to be modern designated two forms of separation.

First, there was a separation or break in time – the contrast between a modern now and a traditional, backward past, so that the societies of the Third World were located in another, previous time and their co-presence in modern time was effectively erased (Fabian 1983). Second, there was a separation in space – a geopolitical distinction made between the modern societies of the West and the 'traditional societies' of Africa, Asia and Latin America. These processes of separation were seen as being accompanied by transformations in science, technology, administration and economy, which were portrayed as opening up the future to limitless advancement and improvement (Adas 1990, Latour 1993). The second kind of separation which reflected the existence of a geopolitical divide was further accentuated by the emergence of the Cold War, which gave a new kind of conflictual significance to the spatial separation between the modern and traditional spheres of world development.

Sakai (1997), in a discussion of the universal and the particular, illuminates significant facets of the connection between the modern/non-modern difference and geopolitics. Thus, while we may wish to keep in mind that the modern has always been opposed to its historical precedent, the pre-modern, geopolitically the modern has been contrasted to the non-West, so that a historical condition is translated into a geopolitical one and vice versa. Although the West is particular in itself, it has been constructed as the universal point of reference through which others are encouraged to recognize themselves as particularities. In the context of post-war modernization theory, the particular universality of the West came to be founded in a process of Americanization, so that as Sakai (1997: 157) puts it, whereas, in the pre-1940 period the process of modernization had been the approximate equivalent of Europeanization, after the Second World War, modernization theory was deployed in a way that reflected the shifting of the centre of geopolitical gravity from Western Europe to the United States.

The emphasis on modernization in the United States can be interpreted as a reflection of a new ethos of national purpose. Emerging at the end of the Second World War as the key Western power, there was a sense within the US that its burgeoning power was the result of the combination of its scientific and technological prowess, its military capacity, its democratic and open traditions and its expansive modernity. The US was the world's number one *modern* nation with a way of life that other less advanced nations would benefit from following and adopting. In its economic, political and cultural spheres, the US was seen as being ahead of other nations, as the nation that could and should offer leadership to the world. Its contemporaneity, rationality, innovation, dynamism, opportunity, mobility and freedom – in a nutshell, its modern being – was a beacon to the world.

But there was another factor which helps to explain the focus on modernization in the context of West/non-West encounters. The term 'modernization' would, according to Walt Rostow, leading economist and deputy national security adviser, replace colonialism and create a 'new post-colonial relationship between the northern and southern halves of the Free World' in which a 'new partnership among free men – rich and poor alike' would emerge (quoted in Latham 2000: 16). Modernization would be conceptualized as a benevolent and universal process, a process based on a certain reading of Western and especially US experience, an experience in which US imperialism was erased, and this process would be framed as being necessary for the modernization and development of non-Western societies, especially relevant in a period in which many of these societies, in particular in Africa and Asia, were undergoing a process of rapid decolonization. The term 'modernization' had a positive, progressive and seductive orientation – to be against modernization would be tantamount to being irrational, backward and retrogressive.

The working out of the will to be modern was exemplified in the writings of a number of American social scientists (see, for example, Lerner 1958 and McLelland 1961). The need for a modernizing will was connected to a negative portrayal of the 'traditional society'. For Watnick (1952: 22), for example, the 'history-less peoples' have a social existence marked by their 'parochial isolation, the fixity of their social structure, their tradition-bound resistance to change, . . . and their static, subsistence economies'. A few years later, Brzezinski (1956: 57–8) referred to the gulf that separated the Western-trained elites from the millions of 'retarded, illiterate, and . . . pliant masses', while subsequently Rostow (1960: 109) wrote of the traditional society as being either incapable of self-organization, or unwilling to organize itself.

These characterizations of the traditional were not unconnected to earlier Western visions of primitive society,[3] and as we saw in the previous chapter, the way the Indian question and US–Mexican relations were contextualized in the US of the nineteenth century provided an initial nucleus of meanings for later interpretations of non-Western peoples. This negative essentialization of the 'traditional society' helped to create an ethos of urgency in which the diffusion of modernization was propelled forward not only to spread the ostensible benefits and achievements of Western and especially American economy and society, but also to help buttress vulnerable regions of the world against the perceived menace of communism.

If the diffusion of modernization was to be successful, the infrastructure for connectivity and change would have to be put in place. Here, the communications media, for example, were seen as instrumental to the modernization process, since their object was not only to open the market to new products and new interests, but also to present the image of a new kind of individual in a new kind of environment (Smith: 1980). Furthermore, within the countries of the developing world, investments had to made in the necessary infrastructure – roads, railways, ports, communications, and public utilities – so that the gulf that was seen as separating the urban, modern centres from the backward, traditional regions could be bridged. The geography of modernization was presented as an exercise in establishing the means for spatially integrating the traditional zones of the developing countries into the national and international circuits of the modernizing world.[4]

While the theory of modernization possessed certain conceptual regularities, nevertheless its orientation did change with the course of events through the years from the early 1950s to the end of the 1960s. In particular, O'Brien (1979), and later Binder (1986) pointed to the shift in emphasis from an earlier concentration on the positive potential of diffusing modern Western democratic ideals, practices and institutional arrangements to a subsequent concern with political order and institutional control. This re-focusing of priorities was particularly manifest in the domain of US political science, where there had been a close connection between the US government and certain influential academics, for example Samuel Huntington (1968), whose specialization related to non-Western areas (Baber 2001; Bilgin & Morton 2002; Gendzier 1985; Mattelart 1994: 153–4). I shall return to this theme below.

What needs to be brought into profile at this point in the argument is that the meanings and practices of modernization could hardly be deciphered without making the connection to the geopolitics of intervention. Such a connection was highlighted by Rostow (1971: 6), who symptomatically suggested that, although the diffusion of modernization was equated with the stage-like trajectory of economic growth, in the last analysis 'the glory of America' has to be related to 'its transcendent *political* mission in reconciling liberty and order' (emphasis added).[5] This bold suggestion provides an appropriate bridge to the next section of the chapter, which, in the setting of the 'three worlds of development', takes up the theme of the geopolitical and historical specificities of the post-Second World War period.

Dynamics of the Geopolitical Context

The geopolitical condition of the world had been profoundly transformed by the end of the Second World War. In the context of the thematic nucleus of this chapter three salient features of the new post-1946 world need to be identified:

a) the emergence of the US as the leader of the Western world, as the pre-eminent hegemonic power;

b) the irruption onto the world stage of the Soviet Union, as an opposing superpower, signalling the beginning of a superpower rivalry that came to mould world politics for a little over four decades; and

c) the emergence of a whole series of new Third World nations, eman-
    ating from the process of decolonization and political independence
    in Africa and Asia; this emergence was combined with related pro-
    cesses of revolutionary change, as in China in 1949 and Cuba in
    1959, and political turbulence in many other non-Western regions –
    in the Middle East, for instance, with the creation of the state of
    Israel in 1948, and especially in Latin America, as, for example, in
    Bolivia in 1952, Brazil in the early 1960s, Colombia in the 1950s, the
    Dominican Republic in the mid-1960s and Peru in 1968.

In 1947, in a seminal foreign-policy statement, President Truman
recognized the new geopolitical situation by stating that it must be the
policy of the US 'to support free peoples who are resisting attempted
subjugation by armed minorities or by outside pressures' (Brockway
1957: 151). Although the immediate context was provided by a request
for aid for Greece and Turkey, countries that were seen as being
threatened by expanding Soviet influence, what came to be known as
the Truman Doctrine made a clear distinction between two ways of life.
One way of life was seen as based on the will of the majority, and
distinguished by freedom, representative government and guarantees of
individual liberty, whereas the other was perceived as based on the will of
the minority forcibly imposed upon the majority, and dependent upon
terror and oppression, a controlled press and radio, and the suppression
of personal freedoms. The drawing of a frontier between two antagonistic ways of life underpinned the transformation of the US relationship
with the Soviet Union from a series of specific conflicts into an overall
geopolitical strategy of global confrontation (Lucas 1996: 281–2).

Two years later, in an inaugural address, President Truman listed four
points in his programme for 'peace and freedom'. The fourth point, as
has been referred to in much of the critical development literature (see for
example Escobar 1995 and Abrahamsen 2000), dealt with the problems
of world development and security. In connecting the 'Point Four Pro-
gram' to the Truman Doctrine there are two elements that can be usefully
identified.

First, poverty was highlighted as a handicap and a *threat* to both the
undeveloped and prosperous areas of the world; and second, the actual
stimulus for the Truman programme on development came, in his
own words, from the need to confront communism (see Arndt 1989:
64). Confronting communism was a key part of a rationale for US
security, and that security received institutional reinforcement in 1947

with the establishment, under the National Security Act, of the Central Intelligence Agency. The so-called 'Fifth Function' of the Agency was interpreted by successive administrations as authorizing covert action, and in 1954 the Doolittle Committee, which had been set up as a Special Study Group by President Eisenhower, argued that an important requirement of national intelligence was an aggressive covert psychological, political and paramilitary organization. The Doolittle Committee Report went on to assert that 'we must ... learn to subvert, sabotage and destroy our enemies by more clever, more sophisticated and more effective methods than those used against us' (quoted in Holden & Zolov 2000: 186–7).

The Truman Doctrine included a new focus on political movements within the Third World. If these movements were judged to be communist in inspiration, they would be actively opposed by the US. In the geopolitical world of the post-1946 years, the primary threat to US security – perceived Soviet expansionism – was not only viewed in global terms but was also depicted as being dangerously mobile and invasive, like a 'fluid stream that moves constantly', attempting to fill 'every crook and cranny available to it in the basin of world power' (Etzold & Gaddis 1978: 86). This new threat was not only global and mobile, it was also envisaged as being intent on penetrating and subverting the vulnerable internal social, economic and political structures of the poor countries of the world.

The overlapping of questions of security, development and modernization and the formulation of global strategy were reflected in a number of statements and interventions from the early 1950s onwards. The belief in a global mission was captured in a statement made in 1952 by the then Director of the Psychological Strategy Board,[6] who suggested that the principles and ideals found in the Declaration of Independence and the Constitution are for export, and are the heritage of everyone – 'we should appeal', he went on, to 'the fundamental urges of all men which I believe are the same for the farmer in Kansas as for the farmer in the Punjab' (qtd in Lucas 1996: 302). The connections between security and the issues of modern development were set in the global context of the US strategy of containment.

The key architect of the strategy of containment was George F. Kennan, and in two highly influential papers he developed an approach which came to form a keystone of US foreign policy.[7] From his 'Moscow Embassy Telegram' of 1946, and his anonymous 'Mr X' paper of 1947, which discussed 'The Sources of Soviet Conduct', three elements can be mentioned.

First, in terms of representation, Kennan's reading of the Soviet Union tended to be orientalist – he wrote of the lessons of Russian history in relation to 'centuries of obscure battles between nomadic forces' and of the nature of the 'Russian or oriental mind', reminding us of the links between the way modernization theorists depicted the 'traditional society' and the earlier orientalist readings of Middle East history and culture (Said 1978). Second, the American necessity of comprehending the new Soviet reality was couched in terms of the way a 'doctor studies an unruly and unreasonable individual', although significantly this 'individual' was now possessed by a new creed – 'communism'. Finally, Kennan's call for a strategy of containment was not only linked to the need for American moral and intellectual leadership, but also to the deployment of a specific geopolitical 'counter-force' (Etzold & Gaddis 1978: 62–3, 87).[8]

Kennan's Euro-Americanist reading of Soviet/Russian history was not a unique example, since in his report on Latin America written in 1950 he presented a wholly negative representation of Latin America and its people, noting for example that in the domain of individuality there was an exaggerated self-centredness and egotism, and that political leaders in Moscow must regard Latin Americans with a 'mixture of amusement, contempt and anxiety' (quoted in Holden & Zoldov 2000: 197). For Kennan, the military significance of Latin America lay largely in its raw materials, essential for the waging of war, and the interests of US security in Latin America required the acceptance of harsh governmental repression if popular governments in the region were unable to resist the communist threat.

Kennan's strategy of containment which was global in scope combined a plea for sustained US hegemony: moral and intellectual leadership, coupled with the capacity to deploy force and repression when appropriate. The interweaving of a hegemonic project, a strategy for containing communism and a negative essentialization of non-Western peoples became a prevalent feature of US geopolitical thinking in the 1950s and 1960s. Modernization and development were in this context constructed as part of an overall strategy to help strengthen the capacity of Third World societies to resist the threat of communist subversion, as well as presenting the West and especially the US as the world leader and diffuser of modern development. Despite the negative portrayal of Third World peoples, it was still believed that with the appropriate mix of investment, reform and foreign assistance, progress could be made and modernization could be diffused, a view that, in the Latin American context, became more prominent in the early 1960s with the Alliance for Progress, discussed below.

However, and re-introducing our theme of the overlapping of the internal and external, the global strategy of containment was not only designed as a principle of foreign policy – it was also a significant motif of US culture during the 1950s (Campbell 1992: 175–82; Nadel 1995). More specifically, as Lutz (1997) convincingly argues, post-war America was characterized by the establishment of a national security state and the beginnings of a militarization process that went together with a new policing of subjectivity. The Cold War was not only a phenomenon of international relations; it went together with the 'Red Scare' and the House Un-American Activities Committee. There was a chilling of debate and a stifling of critical thought in every area of culture inside the US. The activities of Republican senator Joe McCarthy were at the heart of the Red Scare, and 'McCarthyism' (as it came to be called) was very much the Cold War at home where traitors, appeasers and deviants were to be rooted out (see Sherry 1995: 170–7). The importance of psychology for the Cold War abroad was also underlined by President Eisenhower, who in 1952 stated that the aim of the US in the Cold War was not the conquering of territory or subjugation by force, but rather its aim was 'to get the world . . . to believe the truth . . . that Americans want a world at peace, and the means we shall employ to spread this truth are . . . called psychological – . . . "psychological warfare" is the struggle for the minds and wills of men' (qtd in Lucas 1996: 301).

The Cold War therefore was associated with containment and intervention at home and abroad – its dichotomous portrayal of 'us' and 'them', of 'friends' and 'enemies', of threats within and without, wove together questions of security, geopolitics, subjectivity and social change, so that, as today, post-9/11, modernizing or civilizing the non-Western other can be represented as a matter of security and intervention that transcends national boundaries.

Security and interventionism were intertwined with the will to wage war, and, as Stephanson (1998: 82) remarks, for the US, communism was the equivalent of war. The first major test of this belief of US foreign policy came with the Korean War of 1950–3, which constituted the first significant challenge to the Truman Doctrine. After North Korean armies moved southward across the 38th parallel, the US responded by sending troops under the aegis of the United Nations, and the war was stalemated at the 38th parallel until peace negotiations in 1953 restored the original boundary between North and South. An estimated 2 million Koreans were killed (Zinn 1980: 419), and the war left in its wake the world's most militarized peninsula. Not only did the Korean War heighten the geopolitical relevance of peripheral regions in the struggle against communism, but also as Kolko (1988: 54–7) indicates, the process of fighting the war, with all its economic consequences, produced new demands, especially for raw materials; and since many of these raw materials, including strategic minerals, had Third World locations, the perceived significance of the developing world to US interests was intensified.

Pivotal geopolitical interventions soon followed in Iran (1953) and Guatemala (1954). In the early 1950s these two countries had a number of political features in common. In each country, a constitutional regime had emerged where the democratically elected leaders (Mossadegh and Arbenz, respectively) were committed to the extension of the principles of democracy, and to programmes of national reform, which included the nationalization of foreign property, as with the Anglo-Iranian Oil Company in Iran and the United Fruit Company in Guatemala. In both countries, as Smith (1994: 192) reminds us, the US, in the name of confronting communism, covertly organized as CIA operations successful interventions against these reformist governments. In both countries the direct beneficiaries of US action were domestic forces obdurately opposed to redistributive reform and the spread of democracy.[9] In both cases the US rather than being a diffuser of democracy acted as a *terminator* of the democratic process.

In the Guatemalan case, the Arbenz government in 1952 had initiated a programme of land reform which included the expropriation and redistribution of uncultivated or fallow land. This led to a governmental expropriation of well over two-thirds of the 295,000 acres of such lands owned by the United Fruit Company in one region of the Pacific coast (Niess 1990: 149). The company's senior management was closely linked into the Eisenhower administration, as exemplified by the facts that the then Secretary of State, John Foster Dulles, was a company shareholder and long-time legal adviser, while the then CIA director, Allan Dulles, had been a president of the company. More strategically, however, as Cullather (1999) and Immerman (1982) have shown, the land-reform programme was interpreted by the Department of State as a potential opening for the radicalization of Guatemala, within which communist elements could play an increasingly influential role. In fact by 1954 at the Tenth Pan-American Conference in Caracas, Secretary of State Dulles roundly asserted that there was not a single Latin American country that had not been penetrated by international communism. This proposition was then used to help push through a resolution which declared that the domination or control of the political institutions of any American state by the international communist movement would not only constitute a threat to sovereignty and political independence but also to the peace of America. This resolution gave Washington a free hand to carry out unilateral actions against Guatemala under the aegis of the OAS (Organization of American States), which had been founded in 1948.

In a view from the other side, the Guatemalan foreign minister in the Arbenz government commented that the Dulles Resolution of March 1954 represented the 'internationalization of McCarthyism' and the violation of the principle of democracy (quoted in Aguilar 1968: 102). Paradoxically, a Central American society which since 1944 had been pursuing a programme of national reform (including a democratic transfer of governmental power in 1951), diversifying agriculture, improving living standards, legalizing trade unions and peasant organizations, and beginning a land-reform programme – in effect, a strategy of indigenous modernization and development – was considered a threat to American security. In addition, it can be suggested that the threat to the US did not come from communism but from the possibility that a radical, reforming government, independent of the US, could offer a counter-example to other societies of Central America and beyond.

The impact of the US intervention was lasting. As one Latin Americanist puts it, Guatemala has the most regressive fiscal system and the most unequal land-ownership pattern in Latin America, and its army has evolved into an all-powerful mafia, stretching its tentacles into drug-trafficking, kidnapping and smuggling (Gleijeses 1999: xxxii). And, in terms of the human cost, there has been the everyday reality of state terror, death squads, torture, disappearances and executions.

Other US interventions in the fast-changing world of the periphery followed a similar pattern. In Latin America, after the Cuban Revolution of 1959, and the installation of a government that not only broke the internalized power of the US in Cuba, but also allied itself with the Soviet bloc, the US responded with a blockade and an unsuccessful invasion at the Bay of Pigs in 1961. In 1964, a *coup d'état* in Brazil brought a pro-American military to power, and in the following year President Lyndon B. Johnson sent 23,000 US and OAS (Organization of American States) troops to the Dominican Republic to restore order in a civil war situation and help install a conservative regime.

In an Asian context, and also in 1965, the US backed the Suharto military coup in Indonesia, which destroyed the only mass-based political party in that country (the PKI – the Communist Party of Indonesia), with estimates of the dead ranging from 700,000 (Chomsky 1993: 58) to a figure of between 450,000 and 500,000 (Kolko 1988: 182).[10] Also in South-east Asia, the Vietnam War came to constitute the most intense and damaging foreign intervention of the post-war years, bringing in its wake the death of 58,000 US soldiers (Buzzanco 1999: 135) and far-reaching domestic consequences (Sherry 1995: 266–79). After Vietcong guerrillas killed 7 Americans and wounded 109 at a US base in South Vietnam in 1965, President Johnson ordered air strikes against North Vietnam and then announced a US$1 billion aid programme. By the time the Americans accepted defeat and withdrew their troops in 1975, Vietnam had been changed forever, with, for example, the death of an estimated 3 million people out of a total population of 18 million in 1970 (see Kolko 1997: 2).

The examples I have drawn on above are situated in predominantly Latin American contexts, briefly supplemented by two South-east Asian cases, and further examples for Asia and the Middle East can be found in Hahn and Heiss's (2001) text on the US and the Third World since 1945. US interventions in Africa were not as pervasive nor as intense during the 1950s and 1960s. A striking exception was provided by the Congo of the early 1960s (Kolko 1988: 192–3; Worsley 1984: 304), with the then CIA Director cabling a warning that a 'communist takeover of the Congo' would have 'disastrous consequences...for the interests of the free world' (qtd in Blum 1986: 175). The emergence, with Western backing, of a Congolese leader (Mobutu) who was prepared to work with Western capital and with the CIA, and who would be less accessible to the communist-bloc countries, was regarded as an acceptable outcome to Washington. Overall, it was not until the mid-1970s, with the defeat of Portuguese colonialism and the coming to power in Angola, Mozambique and Guinea Bissau of Marxist-inspired regimes, that the US began to assign Africa more strategic significance.[11]

The projection and deployment of US geopolitical power into the societies of the Third World has to be set in an overall context of rapid socio-economic change, intensifying political conflict and Cold War antagonisms. The prevailing sense of acute geopolitical turbulence was accurately portrayed by Hobsbawm (1994), as already mentioned in chapter 1, when he suggested that in general the Third World differed from the societies of the West in one rather fundamental respect, namely that it formed a world-wide zone of revolution. For Hobsbawm very few Third World states of any size passed through the period from 1950 without revolution, military coups to prevent or propel revolutions, or some other form of internal armed conflict. This association of war and conflict with the societies of the periphery during the 1950s and 1960s found another expression in President Kennedy's suggestion in 1961 that the Third World constituted 'the great battleground for the defense and expansion of freedom today' (qtd in McDougall 1997: 184–5). Moreover, the idea that the Third World formed a battleground for the defence and expansion of freedom found institutional expression in the bringing together by the Kennedy administration of the State and Defense Departments, the CIA and the Agency for International Development so that they could analyse on a regular basis what Kolko (1988: 130) reminds us were officially defined as the 'problems of development and internal defense'.

The connections between violence and the spread of modern development are sometimes ignored (for a well-known exception within the modernization school, see Huntington 1968: 40–2), but they have always been interwoven (see for example Shapiro 1997). In a consideration of war and development, the Latin American sociologist Elguea (1991) pointed not only to the predominant concentration of wars in the Third World since the late 1940s, but also to the continuing intersection of violence and development in the histories of Latin American societies. Klein (1994: 98) makes a similar observation, reminding us that the violence at work in the modernization process is not only the strategic-instrumental one of armed forces waiting to intervene in given circumstances, but that violence also occurs at the level of everyday life, as long-standing, collective and communal forms of life are transformed into modern, commodity-based exchange economies.[12]

The relationships among war, political violence, security and modernization were clearly visible in the Washington administration's response to the Cuban Revolution. In March 1961, a month before the unsuccessful US-sponsored invasion of Cuba at the Bay of Pigs, President Kennedy announced his plan for an Alliance for Progress, which was launched with the aim of creating new forms of international cooperation and security among the countries of the western hemisphere. In his speech to the assembled Latin American diplomatic corps in Washington, Kennedy emphasized the common destiny of the US and the Latin American republics, noting, for instance, that 'our nations are the product of a common struggle, the revolt from colonial rule' (qtd in Holden & Zolov 2000: 227). Quoting Bolívar and Benito Juárez, Kennedy stressed the need to combine political freedom with material progress, invoked the guiding example of American civilization, and identified the Cuban government as the key threat to progress and freedom in the hemisphere, while at the same time offering a hand of friendship to the Cuban people.

Through the Alliance for Progress, Washington pledged, for a 10-year period, US$20 billion in public and private funding to help Latin America carry out the social changes thought necessary for the region's progress. US aid to Latin America witnessed a substantial increase, with bilateral economic assistance nearly tripling between 1960 and 1961, and under Kennedy and Johnson, Latin America received nearly 18 per cent of total US aid, compared with just 3 per cent under Truman and 9 per cent under Eisenhower (Smith 2000: 152). The Alliance for Progress was associated with a charter, signed at Punta del Este in Uruguay in August 1961, which established a series of modernizing goals, such as raising per capita income to achieve self-sustaining development, social reform, especially focused on 'unjust structures of land tenure and use', trade diversification, industrialization, and the development of new educational technologies to promote the elimination of illiteracy, a goal posited for other Third World regions at this time (see Mattelart 1994: 158–62).

The ultimate purpose of The Alliance for Progress was to bring together socio-economic modernization with political democracy and in a way that replicated the US model. Although, as Lehmann (1990: 22–4) appropriately suggests, the Alliance placed more emphasis on land reform than had previously been the case, this was partly a response to the geopolitical implications of the Cuban Revolution but also to a growing realization within some sections of the policy-making community in Washington that if modernization as socio-economic transformation was to be successful, a more equal distribution of resources and income was required (Schoultz 1999: 353). However, such a vision also carried with it the danger of a radical process of social mobilization and the coming to power of governments that might not be sympathetic to US interests. Nevertheless, according to Arthur Schlesinger, Jr, White House adviser in the Kennedy administration, Latin America was not only ripe for modernization but also the US had the responsibility to assist in the drastic revision of the 'semi-feudal agrarian structure of [Latin American] society which still prevails' (qtd in Latham 2000: 80). The US had thus to ensure reform of these unequal structures, so that not only would modernization be successful but equally the Latin American republics through such a process would be less vulnerable to the 'communist threat' as illustrated by the Cuban Revolution.

The Alliance for Progress was closely linked into the tenets of modernization theory. Advisers to Kennedy such as Rostow and Schlesinger emphasized the need for the social, economic and political structures of Latin American societies to be reformed and modernized so that these societies could eventually become a southern model of the US itself. Kennedy himself in his 1961 speech underlined the significance of science, industry, knowledge and freedom, and the twin foundation of economic progress and social justice. What is striking about the speech is the way commonalities across the North–South divide of the western hemisphere are invoked, the way an appeal is made for a common struggle against tyranny and disorder with the notion of 'our continents' being bound together by a common history, which is now threatened by 'alien forces' that seek to impose the 'despotisms of the Old World on the people of the New' (Holden & Zolov 2000: 226–8).

The eventual failure of the Alliance for Progress was closely associated with its underlying orientation, which stressed the politics of partnership where in actual fact there existed a top-down relation of hegemony, which exhibited more leadership than persuasion. Latin American governments were regarded as recipients, not to be actively engaged in the working out of the strategy for transformation. Furthermore, reform, and especially land reform, was strongly resisted by the dominant layers of Latin American society, and of more than 15 million peasant families living in Latin America in the early 1960s, Smith (2000: 154) estimates that fewer than 1 million benefited from any kind of agrarian reform policy.

Also, as Schoultz (1987: 19) has pointed out, one has to remember that the Alliance for Progress initiative was rooted in security concerns and Cold War geopolitics, and therefore stability and order were prioritized over social justice and reform. The association of the Alliance with security and order was reflected in the new or expanded counter-insurgency programmes – the Army's Special Forces, the Green Berets – and the Agency for International Development's Office of Public Safety, which was responsible for searching out and destroying communist insurgents. USAID also sent experts and up-to-date equipment designed to counter the 'internal enemy' through a programme of knowledge transfer, in which courses on mass psychology, the 'pathology of insurgents' and the use of photographic techniques in demonstrations were disseminated to other societies (Mattelart 1994: 155).

The Alliance for Progress represented a specific foreign aid initiative that was discursively rooted in key components of modernization theory and was deployed as a response to the geopolitical impact of the Cuban Revolution. Modernization theory itself embraced a concern for transforming peripheral nations into subordinate versions of Western and especially US-style modern societies. At the same time it provided an interpretative and policy-focused frame to confront the threat of radical anti-capitalist movements in a context of the Cold War. In terms of the three worlds of development, the First World, in this case the US, was framed as the universal model, as the leading society whose own modernity was the guiding light for others to follow. The Second World, the Soviet Union and Cuba in particular in the Latin American context, were conceived, due to their posited tyrannical and expansionist nature, as the overriding threat to the diffusion of the American model. The Third World, in this example Latin America, was portrayed as the vulnerable, polarized, inefficient part of the western hemisphere that needed modernizing to save it from itself and the threat of communist subversion. I shall return to this aspect of the argument below.

Although Latin America remained a significant object of US security concerns and a primary zone for the diffusion and deployment of modernization theory, from the late 1950s until the mid-1970s, most US aid was directed to Asia as a corollary of the Korean and Vietnamese Wars, and not surprisingly this aid was mostly directed into outlays for military and security support purposes (Black 1991: 58–63). This aid was also contextualized in relation to modernization ideas. The Korean and Vietnamese situations were part of a larger picture of political turbulence in the Third World, which was anchored in the processes of decolonization, radical nationalisms and the overall challenge to Western power.

In the aftermath of decolonization, newly-emerging nations came to play an increasingly vital role in world politics – as one indicator of the change, membership of the United Nations increased from 51 to 126 between 1945 and 1968. In 1955, the Bandung Conference in Indonesia brought together 29 African and Asian countries to explore their common interests in the struggle against colonialism, and it was here that the 'non-aligned' movement was effectively born, becoming formally established in 1961, one year before the coming together of the Group of 77 (Rist 1997: 82–8). The phenomenon of 'Third Worldism' (Chaliand 1978), as it came to be known, was fuelled by the processes of decolonization and the emergence in a broad range of countries of radical political movements, sometimes referred to as 'national liberation movements', which challenged the power of the West. Further to the surfacing of radical movements, the US also faced the growth of economic nationalism in Latin

America, whereby first in Cuba and then in a range of other countries (Sigmund 1980) the politics of nationalization challenged the historically established practices of multinationals to repatriate profits and make decisions that followed the interests of the company rather than the Latin American economy. US corporations felt this pressure most acutely in the resource-extraction sector, where investments in mining and petroleum had been particularly significant in the post-war period, and where access to strategic raw materials for industrial and military development was seen as a key factor in the evaluation of foreign policy.

Order, Security and Modernization

Having drawn out one particular outline of certain relevant features of the geopolitical setting in which conceptualizations of modernization emerged and developed, it is now necessary to take up in a little more detail the linkages between the changing trajectory of modernization theory and the concern with order, security and war.

The Vietnam War not only had a pivotal impact on US policy at home and abroad; it also influenced the thinking of key US proponents of development strategy for the countries of the turbulent periphery. In the early 1960s, for example, Rostow announced that it was precisely in Vietnam that the liberation war had to be broken; 'if we don't break it here...we shall have to face it again in Thailand, Venezuela, elsewhere...Vietnam is a clear testing ground for our policy in the world' (qtd in McDougall 1997: 187). It was also in the early 1960s that Lucien Pye, a prominent American political scientist, inaugurated, at the suggestion of the Pentagon, a series of studies that were to investigate the potential of the military for guiding economic development in the poor countries of the world.

Similarly, another influential political scientist, writing in the mid-1960s, viewed modernizing societies such as Indonesia or Egypt, Ghana or Tanzania as populist and *pre*-democratic. For Apter (1987: 55), interpreting such societies as pre-democratic 'allows us to view certain institutions of coercion as perhaps necessary to the organization and integration of a modernizing community'. Given that modernization was seen as a complex process of transformation which entailed growing social and economic differentiation and the absorption of externally generated innovations in the fields of science and technology, industrialization, communications, agricultural development, education and

administrative capabilities, it was not surprising that the potential societal stress of modernization was also associated with the dangers of
'breakdown', especially when the presence of ideological ferment was
also taken into consideration.

Much attention was allocated to the role of modernizing elites and
their capacity to engineer Western-led development and change. Eisenstadt (1966), for instance, stressed the centrality of their role in the
crucial problem of developing an institutional structure capable of
absorbing continually changing social problems and demands. Here the
attitude of elites was examined in terms of their relation to modernity,
and institutional patterns were distinguished on a basis of democratic
versus oligarchic, so that with respect to the former, one had a contrast
between the political democracies of the West and the tutelary or guided
democracies of the non-West, and as regards oligarchies, it was possible
to differentiate three types: the modernizing, as in those peripheral
societies that were moving away from being traditional and backward;
the totalitarian, as in the countries of the communist world; and the
traditional, as in those countries that had still not embraced the diffusion
of Western modernity. In these kinds of formulations, modernization
always had its shadow – the danger that through the process of social
mobilization, and as an inevitable component of the 'revolution of rising
expectations', new strata or groupings would emerge that would be
hostile to the Western-led process of modern transformation. Infused
with the extraneous ideology of communism, these radicalized layers of
the modernizing society could well come to pose a serious and sustained
threat to an orderly transition to modernity. It was in this context, for
example, that Pool argued that 'order depends on somehow compelling
newly mobilized strata to return to a measure of passivity and defeatism
from which they have been aroused by the process of modernisation'
(qtd in Higgott 1983: 19).[13]

One of the most influential contributions to the growing concentration
on issues of order and institutional stability was provided by Huntington
(1968), in a text which came to be seen as a central intervention in
the debates on modernization and political conflict in changing societies.
Huntington began his analysis by drawing attention to the increase in
political disorder during the 1950s and 1960s. He noted, for instance,
that violence and other destabilizing events were five times more fre-
quent between 1955 and 1962 than they were between 1948 and 1954,
and that while in 1958 there were 28 prolonged guerrilla insurgencies, 4
military uprisings and 2 conventional wars, by 1965 these figures had

increased to 42, 10 and 5 respectively. It was argued that throughout Asia, Africa and Latin America there was a decline in political order, and an undermining of the authority, effectiveness and legitimacy of government. (Huntington 1968: 4). For Huntington the countries of the Third World lacked civic morale, public spirit and the stabilizing influence of mature political institutions.

The essential problem of politics was the failure of political institutionalization to keep up with the pace of social and economic change. The crucial problem for developing societies was one of order, authority and institutional control, features that in the US had been assured since its founding.

For Huntington it was not the absence of modernity but the effort to achieve it that produced political disorder, and it was actually the diffusion of modernization throughout the world which increased the prevalence of violence (Huntington 1968: 41). While he associated violence with Third World society itself, rather than with the nature of Western power, he did point to the stabilizing effect of revolutionary governments which did secure order while introducing totalitarian rule. Organization, for Huntington, was of cardinal importance, and in this he credited Leninism with bringing to modern politics not the destruction of established institutions but the organization and creation of new political institutions. In general, it was concluded that political organization was crucial for stability and liberty – 'in the modernizing world he controls the future who organizes its politics' (1968: 461).[14]

The trajectory of modernization theory was greatly affected by the intensified geopolitical turbulence in the societies of the periphery. The concern with order and institutional control, made so much more palpable by the Cuban Revolution and the Vietnam War, came to receive more attention in the 1960s, and the Huntington text was emblematic of this shift in focus. Such a shift in analytical attention was not only due to greater turmoil in the international arena, but was also influenced by the development of social conflict within the US itself. Opposition to the Vietnam War which developed through the 1960s and into the early 1970s – especially visible on US campuses with the activities of Students for a Democratic Society, in the upsurge in the civil rights movement, and in the destabilizing effect of assassinations of key leaders such as Martin Luther King Jr and the Kennedys – tended to create an ethos in which a concern with order became more prominent. Overall, the earlier, more optimistic visions of the gradual but effective diffusion of modernization and democracy came to be overshadowed by a focus on political conflict and the need for authority and order to help buttress Western power in a destabilizing world. This change in focus evidently reflected the dynamic of Cold War politics and the intensification of superpower rivalry from the late 1950s through to the mid-1970s and the end of the Vietnam War.

Clearly, the concern for order and stability was not entirely new in the sense that in earlier periods of the twentieth century similar preoccupations had been voiced, as we discussed in the previous chapter. It provided another example of the drive to 'pacify the political' in the sense employed by Rancière (1995), and discussed previously: i.e. the desire to impose a settled order on a series of disrupting and destabilizing currents and forces present within a society at a specific moment of time. In the particular case of the perception of world order in the 1960s, the difference with the earlier era stemmed from the fact that the prioritization of order, institutional stability and security was formulated in a radically altered geopolitical world by a combination of US social scientists and government advisers specializing in and concerned with the problems of rapid change in non-Western areas. Paradoxically the more the drive towards Western-style modernization was encouraged, the more likely it became that threats to institutionalized orders would emerge and endanger the stability of societies undergoing processes of social, economic and political transformation.

Revisiting the Limitations of Modernization Theory

In marshalling together the main elements of my critical assessment of the theory of modernization, I want to focus my attention on three basic considerations. If we accept the heuristic validity of the Three Worlds distinction for the period under review (Pletsch 1981), i.e. for the 1950s and 1960s, and work from the assumption that there were, and perhaps still are, in sharply modified form, some worthwhile aspects of such a distinction, then we might usefully begin with the so-called First World.

Predominantly, critiques of modernization theory have tended to remain silent on the way the First World itself was situated and interpreted. This contrasts with the relations between First and Third Worlds, or the actual representation of the Third World itself, which have both been the subject of much critical discussion. My objective here is to introduce one or two new elements and recover other facets of older critiques which have tended to be forgotten, and which are relevant to the following chapter's examination of neo-liberal approaches to being modern.

Representations of the First World

O'Brien (1979: 53–4), in a strikingly prescient observation which anticipated Fukuyama's (1992) 'end of history' thesis, suggested that the idealized version of being modern, as projected by the theorists of modernization, had an American face, and that 'this ideal type is in effect the end of history, the terminal station at which the passengers to modernization can finally get out and stretch their legs'. Idealized visions of the West, with the US as its prime mover, were certainly a key part of modernization theory, and since such a tendency is distinctly present in contemporary neo-liberal portrayals of Western modernity, it is worthwhile reflecting on this issue in a little more detail than is normally the case. What I would argue here is that the darker side of First World history was excluded from the dominant discourse of Western modernization. The development of West European and North American societies was described in a language free of substantive criticism, and certain rather crucial social, economic and political phenomena were left out of account. This was not an attribute confined only to the theorists of modernization, and nor was it to fade away with the demise of this particular regime of truth, as will be noted in the following chapter.

Thinking about the dark side of Occidental history, it was the Argentinian social scientist Bagú (1970: 60–1) who drew attention not only to the importance inside the countries of the West of the systematic exercise of violence on a large scale – the phenomena of robber barons, organized crime and corruption – but also to the fact that such revealing social realities tended to be ignored by social and political theorists. This was also the case with more critical writers such as C. Wright Mills (1956), who in his book on *The Power Elite* analysed other trends relating to violence and the power of the military in American society, but not the mafia and gangsterism.[15]

Organized crime, violence and corruption were and of course still are salient features of American society, as elsewhere, but their place and meaning within theories of modernization seemed to have had little if any relevance. Were these features only to be considered as extrinsic to the drive to become modern? If they were not part of the process of modernization, how were they to be situated – as irrationalities of the transition from pre-modern to modern? Was the diffusion of modernization to the countries of the Third World to include the spread of well-organized, well-funded, achievement-oriented crime syndicates? In reality that sometimes happened, as was the case in 1950s Cuba where pre-1959 dictator Fulgencio Batista provided a safe haven for US mafia investments in gambling casinos and Havana's tourist industry (Pearce 1978: 150–1 and Pérez 1997: 222–5). Although statistical information on organized crime has always been difficult to obtain, Pearce, in his interesting but neglected study of the 'crimes of the powerful', notes that according to the President's Commission on Law Enforcement in 1967 it was estimated that in the US organized crime made a profit of US$7 billion from gambling alone, and two years earlier it had been estimated that the annual profits of organized crime were of the order of $9.5 billion (Pearce 1978: 77–9). Placing such a figure in another context, it might be compared with US direct private investments in Latin America, which in the same year, 1965, were at $9.4 billion.[16]

The political histories of First World societies were also presented in a way that tended to make invisible a range of social phenomena that were quite fundamental to any effective understanding of such societies. In the specific case of the US, one might mention: the nineteenth-century history of territorial expansionism, including the violence of the movement west, referred to in chapter 2 above, the War with Mexico in the 1840s, the relevance of the Civil War of the 1860s, and the continuing rootedness of racism in the southern states of the Union and more generally (see MacCarthy 2002); the influence of race on the design and implementation of foreign policy towards other less powerful societies in the Americas and elsewhere (again touched on in chapter 2); the impact and meaning of military interventions with the attendant creation of semi-protectorates, or in the example of the Philippines, colonial annexation, and more contemporaneously for the modernization theorists of the 1950s and 1960s, the political intolerance of 'McCarthyism'.

For Western Europe, the history of the slave trade and the violence associated with the later annexation of overseas territory and the establishment of Empires; the irruption of fascist movements in the 1920s and 1930s (Mazower 1999); and the exercise of colonial violence by Britain, France and Portugal in the post-war years were all intrinsic to the development of Western modernity, but generally absent from the dominant contextualizations of the modernization process. Essentially, modernization was constructed as a process liberated from the destabilizing influence of these other phenomena, which functioned as a shadowy outside. In this sense there was a crucial split at the heart of the theory, whereby the representation of the West went together with the erasure of some of the most disturbing and disruptive aspects of its inner structures and dispositions. Symptomatically, many of these repressed features – for example, violence, corruption, and disorder – were then externalized and projected on to the non-Western other, thus helping to stimulate the desire to penetrate, police and control, while at the same time validating a partialized, narcissistic Western identity.

Portrayals of relational West versus non-West dynamics

As is well-known, one of the central tenets of modernization theory was that the interaction and evolving interrelationships between the modern and traditional would become a mutually beneficial process.[17] In particular, the traditional society, through absorbing the Western diffusion of innovations – capital, technology, entrepreneurship, modern institutions and values and the spirit of achievement – would go through a process of adaptive 'upgrading' which would include economic take-off to industrialization, democratization via Western law and science and rationalization through education.[18] As a critical response to this kind of proposition, it was clearly shown that in the economic context the relations between the countries of the First World and those of the Third exhibited an inequality that contradicted the sanguine predictions of modernization theory (Taylor 1979: 42–70).[19] Furthermore, it was convincingly argued by a range of authors that the relations between developed and underdeveloped countries could not be understood outside the context of power relations and the negative impact of Western invasiveness. Also, with regard to the diffusion of values and institutional models from the First World, it was countered that the imitation of institutional structures was dysfunctional for the Third World society (Whyte 1968). This view, as we shall see in subsequent chapters, was re-asserted and re-invigorated in future years.

Essentializations of the Third World

The final criticism here relates to the ways in which the societies of the Third World were characterized. We have already mentioned the construction within modernization theory of a negative essentialization of the traditional, and its link to Western notions of the 'primitive'. At the same time, as Said (1981: 28–9) pointed out in his discussion of Western interpretations of Islam, the mainstream literature on modernization bore a close resemblance to orientalist preconceptions. Said suggested that one of the illusions of modernization theory was the notion that before the coming of the US, 'Islam existed in a kind of timeless childhood, shielded from true development by an archaic set of superstitions, prevented by its strange priests and scribes from moving out of the Middle Ages into the modern world' (ibid.). The link here to Said's critique of orientalism is important since in the subsequent development of neo-liberal ideas, as well as in other more critical domains of Western thought, representations of the non-Western other have retained some of the features of earlier modernization theory. A key part of the critique of modernization related to the stress on the heterogeneity of Third World societies, and on the complexities and differences in their social, cultural and political histories.

As one example, it is instructive to recall that in Latin America projects of modernization had already been underway before the era of modernization theory. In the Peru of the 1920s, President Leguía initiated a programme of modernizing infrastructure and public services together with a strengthening of the export economy that he likened to a project of the Americanization of Peru (Carey 1964). In other countries such as Brazil, Mexico and Chile, internal elites had similarly begun programmes of modernizing their economies through state investments in infrastructure, public utilities and industrial development (O'Brien 1996). Such programmes continued during the 1950s and 1960s, as Coronil (1997) shows for the Venezuelan case. Therefore, for the Latin American domain of the Third World, modernization was not just an externally-diffused phenomenon, tied to Western capital and innovations; it had its own internal history and dynamic.

Finally, a further feature of the rather rigid portrayals of the Third World concerned the issue of vulnerability – the notion intrinsically embedded in Cold War geopolitics that the societies of the periphery needed to be protected from an expansionist, penetrating ideology of communism that had the potential to undermine and subvert the structures of these societies. Movements for radical change were thus frequently interpreted as the effect of outside penetration rather than the authentic expressions of independent internal social forces, the Guatemalan case being a pertinent example.

My perspective on the pitfalls of modernization theory is linked into the post-colonial view that it is necessary to identify three interrelated components: the representation of the First World itself, portrayals of the relations between the 'modern' and the 'traditional', or First and Third Worlds, and the representation of the Third World itself. But this is not all, since there are other features of the post-colonial that are relevant to a summarizing of this chapter's orientation.

Theory, Memory and the Post-Colonial

I have indicated how events and trends inside the Third World, and most notably the Cuban Revolution and the Vietnam War, but more generally the overall geopolitical turbulence of the societies of the South in the period under review, impacted on the evolution of Western ideas on modernization. The Third World was not a passive screen for the projections of Western theory. It was an active, dynamic part of the post-Second World War world that greatly affected the nature and trajectory of the theory of modernization, especially evident in the shift of conceptual focus towards a greater emphasis on questions of order and security from the early 1960s onward. In addition, the conceptual terrain of modernization was mapped out in a context of Cold War challenges that transcended the internal/external divide. Domestic politics in the US were as much influenced by superpower rivalry as international politics, and part of a post-colonial perspective would give extra emphasis to this imbrication of inside and outside.

Modernization theory which stressed the primacy of transformation had a multi-dimensionality that gave it a scope and depth that was rather unique. It was not just about the stages of economic growth and a notion of 'take-off' – rather, it prescribed certain kinds of change for the social, psychological, political and cultural dimensions of non-Western societies. In a time of rapid change and instability, it was closely associated with key geopolitical interventions in Third World societies, with the further extension of imperialist penetration, lending such interventions a conceptual and thematic legitimization. It offered a more systematic, transformative approach than had the pre-1940 themes of civilization, order and progress, although it built on those earlier perspectives and retained and re-articulated their central meanings. It provided a codification of what it might mean to be 'modern' as opposed to 'traditional', and of particular significance for the orientation of this study, modernization theory attempted to provide an interpretation of some of the central issues of the times through taking into account the gravity of Third World politics. It was certainly universalist and ethnocentric or Euro-Americanist, based in a view of the singularity of Western modernity,[20] but it was not narrowly econocentric in purpose, and there was a role for the public sector in development and change. Not all matters of modernization were assumed to be best taken care of when left to the private sector. In this sense there was an important difference with the later working out of neo-liberal approaches to development and change, which, as will be argued in the next chapter, have acutely prioritized the role of the private sector, re-asserting commodification and competition.

One of the singularities of modernization theory, and one of the differences with neo-liberalism, was that the nation-state was given more significance in the prosecution of economic development. It is true, as Munck (2000: 37) has reminded us, that the relevance given by modernization theorists such as Rostow to economic growth has received continued weight, as in the 1987 Brundtland Report; but the theorists of modernization were advocates of a public/private sector articulation that was more balanced than that of the neo-liberal persuasion. The nation-state still remained a pivot of geopolitics, prior to the advent of structural adjustment, monetarism, privatization and the intensified commodification of social life.[21]

What the protagonists of modernization share with later theorists is a firm belief in the superiority of the West and the ostensibly beneficial impact of the Western diffusion of capital, institutions, democracy, achievement, rationality and (now again) civilization.[22] What the advocates of modernization also share with many of today's politicians, writers and observers is a continuing belief in the right of the West, in specific circumstances, to impose its geopolitical will on the ostensibly recalcitrant non-Western other. A critical post-colonial sensibility can help remind us that the current re-assertion of imperialist viewpoints has a long history which takes in the essential postulates of modernization. Far from being an innocent or neutral or objective discourse of how a society might become modern, modernization theory was part of the conceptual architecture of a diffusing imperialist logic. Remembering and re-analysing such connections can help reinvigorate a geopolitics of counter-memory, so that one focuses not only on events and circumstances, nor only on ideas and concepts, but on their overlappings and intersections.

4

The Rise of Neo-liberalism and the Expansion of Western Power

'Even those who wished most ardently to free the state from all necessary duties, and whose whole philosophy demanded the restriction of state activities, could not but entrust the self-same state with the new powers, organs, and instruments required for the establishment of *laissez-faire*'
— Karl Polanyi (1957: 140–1)

'[T]he West has been and remains victorious – and not only through the force of its weapons: it remains so through its "models" of growth and development, through the statist and other structures which, having been created by it, are today adopted everywhere.'
— Cornelius Castoriadis (1991: 200–1)

Introduction

By the early 1970s, modernization theory had gone into decline, subject as it was to widespread critique, especially from social scientists in Latin America (as will be seen in chapter 5). Moreover, the defeat of the US in Vietnam accentuated the tendency to seek out less-overarching interpretations of North–South relations. Also, concern with declining rates of profit in the economies of the West helped to stimulate an emphasis on monetarist policy, especially pioneered by the IMF, with the re-assertion of the importance of the private sector and the need for more support for capitalist enterprise. With the advent to power of Ronald Reagan and Margaret Thatcher, the early 1980s were witness to the beginning of what became known as neo-liberal policies on social and economic change, with privatization, deregulation and a championing of market forces being key themes. The neo-liberal approach to development and change also came to be applied by the World Bank, and from the beginning of the 1980s policies of structural adjustment and privatization came to be increasingly influential, especially after the Mexican debt crisis of 1982. The 1980s were characterized by a twin emphasis on the salience of the market and Western-style democracy. By the end of the decade in 1989, with the collapse of the Berlin Wall, neo-liberal ideas were extended from the economy to include notions of 'good governance' and refashioned conceptualizations of the state.

Although neo-liberalism has been universally deployed, I intend, in keeping with my overall analysis, to concentrate on the relevance of neo-liberal ideas for North–South relations, and in so doing I will situate these ideas in an historical as well as social and political context. My main objective is to locate the neo-liberal discourse of development in a frame that is not limited to a social-economic analysis, but rather encompasses broader geopolitical and historical issues. In this context, the chapter also includes a short consideration of new forms of geopolitical interventionism under the rubric of the Reagan Doctrine which anticipated some features of the current foreign policy of the Bush administration, to be discussed in chapter 7.

Not only is it important to understand the genealogy of neo-liberalism, including an understanding of the place of the 'neo', but also it is necessary to be aware of changes of focus and interpretation within the 'regime of truth' that is neo-liberalism. Equally it is important to show how these alterations of orientation (for example, from 'structural adjustment' to 'good governance' and to 'social capital') are themselves a response both to changes in world geopolitical relations and the unfolding dynamic of North–South linkages.

In this chapter, I intend to argue the following : (i) that the deployment of neo-liberal ideas on international development and social change has been part and parcel of an enlargement of occidental power in the period since the early 1980s; (ii) that while the roots of today's neo-liberal globalization lie in the late 1970s and 1980s, equally, neo-liberalism has a longer history than is sometimes realized; (iii) that the neo-liberal discourse of development has a dynamic that needs to be understood; (iv) that given the global hegemony of neo-liberal ideas, it is clearly necessary to critically consider the way these ideas have been formulated, disseminated and broadened in the context of North–South relations: and (v) that in the case of US foreign policy in the 1980s, the projection of market-based or neo-liberal democracy was part and parcel of the Reagan Doctrine of confronting or 'rolling back' communism.

Within this framework, a post-colonial perspective highlights the need to understand the specificity of the second major wave of mainstream Western theory since the Second World War, and especially its impact on the societies of the periphery. Equally, it has to be taken into account that within the South itself, neo-liberal ideas have been developed *sui generis*, as the Chilean case post-1973 shows, and have also varied considerably in their application within the periphery, depending on the specificities of internal politics and social structures (see e.g. Demmers, Fernández Jilberto & Hogenboom 2001).

Designing Development for the 1980s

The 1980s was a decade of 'structural adjustments', of the imposition of a new era of financial discipline to solve the rising debt problem, and of the streamlining of the state. The effects of structural adjustment programmes are both well known and well debated. In the 1980s they included the devaluation of local currencies, a trend towards the dollarization of the Third World economy, increased producer prices and reduced wage bills, wage freezes, a decline in the level of real wages and salaries, the reduction of government subsidies and tariffs, the diminution of government expenditure on welfare and social services, and the privatization of state enterprises. The driving concepts were to cut, to discipline, to differentiate and to restructure, but 'adjusting the structure' was far more than an exercise in economics. To illustrate this suggestion, we may refer to the Bolivian case.

In 1985 a freshly elected government launched a New Economic Policy, which was anchored in the belief that while the state was a key obstacle to successful economic development, free trade, an open economy and market forces constituted the long-term solution. The new economic policy was implemented as part of a far-reaching political intervention.[1] Following enactment of the governmental decree of August 1985, a 'state of siege' was declared and hundreds of COB (*Central Obrera Boliviana*) leaders were sent into 'internal exile' (Conaghan, Malloy & Abugattas 1990: 25). Suspension of constitutional guarantees and rights for political and trade union leaders were accompanied by the violent repression of demonstrations. At the same time, the government 'decentralized' the State Mining Corporation into regional enterprises with the dismissal of 23,000 workers (Dunkerley 1990). Overall, urban unemployment rose from 5.8 per cent in 1985 to 10.2 per cent in 1989 (ECLAC 1989: 18), while underemployment in the main urban areas was estimated to have risen from 62 per cent in 1987 to 73 per cent in 1990 (ILDIS-CEDLA 1994: 86). The main economic measures of the new strategy included the devaluation of the currency, the removal of restrictions on imports and exports, the freezing of public sector wages for 4 and later 3 months, abolition of the minimum wage, and an end to fixed prices on most goods and services. In 1987, as a way of mitigating the negative repercussions of Bolivia's structural adjustment package, an emergency social fund was established with the support of the international financial institutions, and most of the foreign resources ear-marked for the fund were channelled through the World Bank (Graham 1992: 1234).

The Bolivian government linked the idea of saving 'la Patria' with the need for private accumulation, connected the 'rational necessity' of 'administering scarce resources' to the policies of monetarism, and stressed the imperative of competitiveness in the context of the 'self-regulation' of the market. Furthermore, as Torrico (1990) has perceptively suggested, the government combined simulation with dissimulation; it simulated the idea of a 'Bolivian miracle', based on the fall of inflation, but strategically ignored the financial incorporation of coca-dollars,[2] and it attempted to dissimulate and camouflage the real social costs by projecting an image of societal acceptance of official policies.[3] By the early 1990s, it was estimated that in a country where over two-thirds of the population were living on no more than one dollar a day, 51 per cent of urban households and 94 per cent of rural households had unsatisfied basic needs. In other words, half of all urban households (approximately 1.5 million people) and almost all rural families (roughly 4 million people) lacked adequate access to drinking water, sewerage, education and health services (Muñoz 2001: 85).

In the Latin America of the 1980s and 1990s, the Bolivian case has been one of the most striking examples of the adoption of a neo-liberal strategy, but from the Chilean experience post-1973 through to Peru and Argentina in the early 1990s there has clearly been a 'silent revolution' in economic and social policies throughout Latin America,[4] with the generalized adoption of neo-liberal policies.

In the African context, Hoogvelt (1987) provided a brief but incisive commentary on the negative effects of IMF and World Bank policies on social stability in Sierra Leone, while Mamdani (1994) was similarly critical of IMF policies in Uganda, and in a specific treatment of the debt problem, Watkins (1994) developed a related argument for Africa as a whole. In an Asian context, Broad (1990) critically evaluated the impact of World Bank and IMF policies on income distribution in the Philippines, and in a broader exploration Biersteker (1990) argued that IMF and World Bank prescriptions undermined the fiscal basis of the peripheral state, and in the process were endangering its long-term legitimacy. Finally, we may note Payer's (1991) study of the politics of debt, which reminds us of the difference in the favourable US treatment of Mobutu's Zaire (now the Democratic Republic of the Congo) compared to Nicaragua during the early 1980s. Payer also alerts us to the persistence of financial dependence, noting that from 1982 to the end of the 1980s, the Third World had been a net exporter of hard currency to the developed countries.[5]

The 1980s began with the World Bank underlining the centrality of economic growth. It was indeed asserted that there was sufficient evidence to posit that 'economic growth generally contributes to the alleviation of poverty', and that in a more general sense, 'human development depends on economic growth to provide the resources for expanding productive employment and basic services' (World Bank 1981a: 67, 97). By the beginning of the 1990s, it was suggested that development was only likely to be successful if it was 'market friendly', and after a decade of 'structural adjustment' the World Bank defined the most appropriate role for government as being one of supporting rather than supplanting competitive markets (World Bank 1991: 11). The role of government was defined in relation to a number of interlocking functions, which encompassed investment in education, health, nutrition, family planning and poverty alleviation, as well as in infrastructure, the mobilization of resources to finance public expenditures, the protection of the environment and the provision of a stable macro-economic foundation. Looking back on the 1980s, the Bank, in its Annual Report for 1992, took the view that 'adjustment policies help most poor people – at least in the medium term', although it was acknowledged that economic reform programmes could cause 'temporary welfare declines for some' (World Bank 1992a: 69). In the Development Report of the same year, it was declared that there was now 'near unanimity on the central importance of markets and human resource investments for successful development' (World Bank 1992b: 178).

In a concomitant fashion, the Inter-American Development Bank (IDB), which has responsibilities for Latin America, published a report in 1991 which replicated the World Bank approach. Four 'strategic directions' for future change and reform were outlined:

a) the encouragement of an outward economic orientation with growing hemispheric integration;
b) modernization through private-sector development, with the state playing the role of closing down or privatizing most of its public enterprises;
c) reducing the size of government, with cutbacks on public expenditures and the creation of a minimalist state allied to an overall 'deregulation and debureaucratization of the economy' (IDB 1991: 14); and
d) the prioritization of 'human resource development', with improvements in health and education being singled out for primary attention. In this context it was indicated that the long term solution to the poverty problem lay in the improvement of professional skills which would allow for an increase in the numbers of people participating in economic development.[6]

As a final example of the uniformity of ideas and prescriptions present from the 1980s until the early 1990s, one may refer to the OECD report on development cooperation for 1992, in which advocacy of deregulation, the provision of an 'enabling environment' for the private sector and of 'macroeconomic stability' punctuated the text (OECD 1992: 20).

What was being clearly expressed in the above passages was the revival of an economic liberalism that prioritized market-orientated development strategies, a minimal state, free trade, financial discipline, and progress and prosperity through economic growth. Strong states, and here one has a clear contrast with modernization theory, were associated with inefficiency, waste, corruption and the incubus of centralized bureaucracies. In contrast, a streamlined state, effectively nurturing an 'enabling environment' for private enterprise, providing social services and training new generations of human capital, was envisaged as the most appropriate aid to an expanding and creative private sector. Furthermore, permeating diagnoses of development, of public/private sector relations, of welfare and poverty, of science and education, of trade balances and financial flows, and of recommendations for governmental policy, one can locate a firmly anchored belief that the 'economic' has been purified of the political. Relevant market mechanisms and rationally operating individuals, dynamic entrepreneurs and efficient international investors, sound money and sensible economic stabilizations are constituted as the essential building blocks for the official language of development. The overall objective is to place a set of contestable constructs under the control of a settled system of understandings and priorities – to implant and institutionalize in a global setting a shared vision of development as best forged through market-based policies.

The statements, modes of classification and guiding meanings are characterized by what may be called a 'politics of amnesia', which is pivotal for the construction of a new truth. From the early 1980s onwards, in the development discourse it is as if the societies of the South have never experienced previous waves of capitalist penetration and modernization, as if their economies have never been open to the world market, as if the post-Second World War diffusion of modernization theory had never occurred. It is as if real development is set to begin with 'structural adjustment' and the liberalization and privatization of the economy. Further, an image has been constructed that all that went before structural adjustment was fundamentally detrimental to the body economic of the developing countries. The existence of a profound malaise, most obviously embodied in the debt crisis, required a long-term strategy for cure, necessitating where appropriate shock-therapy, rehabilitation, infusions, donors, special treatment for debt distress, relief measures, support against adjustment fatigue and, crucially, continual monitoring and surveillance. The notion of a politics of amnesia, which in this case concerns the history of North–South relations and the omission of past waves of modernization and the insertion of peripheral economies into the international system, is also relevant in the context of an association between the economy and medical metaphors, as the existence of previous 'money doctors' during the 1920s bears witness (Drake 1994).

What we need to do at this stage of the argument therefore is to trace out some of the historical continuities in the contemporary deployment of neo-liberal perspectives so that we can better situate and understand this particular representation of social and economic change. Current assumptions concerning the self-regulation of the market, the unquestioned status of private property, a minimal (and subsequently effective) role for the state, the cultural superiority of the West, the positioning of the individual as possessor, and the centrality of acquisitiveness have a deeply rooted history.

Traces of a Tradition: Problematizing the 'Neo' in Neo-liberalism

Writing in the early 1940s, and in the context of the influence of eco-nomic liberalism, Laski (1946) argued that unregulated competition had never resulted in a well-ordered society, and unless the acquisitive im-pulse could be harnessed to an agreed social purpose, society would always be riven by discontent. In relation to the dominant assumptions of the day – that private property in the means of production was regarded as sacred; that whatever was to be done by government could be better done by private enterprise; that the law is designed and enacted in the common interest, and that each individual knows his own interest best – Laski concluded that the results of economic liberalism had been a grim failure for the majority of the world's population (Laski 1946: 15).[7]

Along a related pathway of enquiry, Karl Polanyi (1957) stated that contrary to the myth of the natural emergence of a self-regulating market, the market was the result of a 'conscious and often violent intervention on the part of government which imposed market organiza-tion on society for non-economic ends' (1957: 250). Such a view is still germane today, and in the interventions that have been referred to as exercises in 'structural adjustment', with debt reservicing and condition-alities, the role of government has been performed by international insti-tutions vested with the power to re-order the economic and social policies of Third World governments. An overriding belief in what Foucault once termed the 'benign opacity' of economic processes[8] has certainly informed the actions and interventions of institutional bodies such as the IMF and the World Bank, but the doctrine deployed can never be fully apprehended if only viewed within the specific compass of 'economic policy'.

Market relations, as a case in point, can be envisaged together with conceptions of the individual. In much modern thinking on the relations between the market and the individual, the image of the individual mirrored the image of market man, and men were portrayed as self-interested calculating machines, not unlike contemporary notions of 'rational man'. If society was a series of relations among the owners of property, political society was a contractual device for the protection of proprietors and the orderly regulation of their relations. In this light, individualism was portrayed as 'possessive individualism' (see Macpherson 1988).

In a similar context, Hirschman (1981) suggested that the pursuit of gain and the self-improvement of the individual were bestowed with a quality of virtuousness that could be traced back through a wide variety of writers from Adam Smith to Keynes.[9] However, the idea that men pursuing their interests would be forever harmless was abandoned when, as Hirschman (1981: 126) puts it, 'the reality of capitalist development was in full view', with its negative effects of increased social polarization and large-scale unemployment.

The primacy of property and the possessive individual, the centrality of acquisitiveness, and the supremacy of the market, as orientating tenets, continued to have an essential place within the tradition of economic liberalism through to the nineteenth century and beyond. They were clearly challenged by the development of Marxist thought and subsequently modified, as the twentieth-century rise of social democracy, with its support for an enlarged socio-economic and redistributive role for the state, clearly demonstrates. The rise of social democracy and the influence of Keynesian-type interventionism in the economy, together with the post-Second World War emergence of welfare states in Western Europe, effectively curtailed the dominance of 'laissez-faire capitalism', and as was shown in the previous chapter, modernization theorists advocated a more substantial role for the state than was the case with the neo-liberalism of the 1980s.

In the domain of international development, neo-liberal ideas have largely emanated from Washington – reflected, for instance, in the 'Washington consensus' on the market economy and Western democracy (see Williamson 1993; and for a critique, Dezalay & Garth 1998). A symptomatic feature of the re-assertion of liberalist ideas has been adherence to a laissez-faire perspective on the economy, and clearly such a perspective has an historical tradition that would encourage us to question the 'neo' in neo-liberalism. The essence of laissez-faire has been the belief that the state must be restrained from undue interference in the economic life of society, and that only through private-sector autonomy, sanctioned by law, can economic activity prosper, society be healthy and individuals realize their maximum potential.

In the US, laissez-faire has rested on a series of interrelated beliefs that are still effective today. From the nineteenth century onwards the American experience of laissez-faire has been connected to a strong sense of natural law which combined the image of a vast continent on which it was founded with, as was suggested in chapter 2, a self-reliant, rugged individualism that was itself seen as part of the order of Nature. The current of rugged individualism that flowed through a nation tied to the significance of the frontier was reinforced by a complementary sense of moral purpose and mission that originated in the Protestant ethic, with its privileging of individual self-interest and the economic virtues of thrift and efficiency. A tradition of moral individualism which came down from Protestant thinkers and from the philosophy of Adam Smith's day, became deeply embedded in the American consciousness. Further to these two intersecting currents of individualism and economic probity, a third component of laissez-faire in the US has been based on a distrust of the state, inherited from the fear of governmental tyranny and European dynastic wars during the time when migration to America began.

Finally, as Max Lerner (1964) suggested in his comments on the anatomy of laissez-faire, there was also a stress on the posited efficacy of competition, as a carry-over from the Darwinian notion of conflict and competition in nature. In an intellectual climate which emphasized natural selection and the survival of the fittest, the allure of success through capitalist competition encouraged an association between the pursuit of profit and an order of Nature – a link, in fact, that Marx had famously made one of the chief targets of his critique of bourgeois political economy. In a climate of this kind, the economists who defended the virtues of capitalist competition became rather more than economists. In legitimizing the given order they became the transcribers of the truths of capitalist progress and competition, and one of their key functions was to expunge the political from the domain of the economic.

Although the influence of laissez-faire ideas has been most notable in the US, their overall impact has been of a global nature. Their retention and redeployment in the late twentieth century are an important part of the genealogy of the neo-liberal. Neo-liberalism in development discourse must not however be seen as static – its content and deployment in the contemporary era have been affected by a series of social changes and mutations of perspective, which will be discussed in the next section of the chapter.

Overall, the above arguments lead me to underline the following three points.

First of all, the contemporary neo-liberal orthodoxy, with its ensemble of market, private property, streamlined state and acquisitive individual, expresses an historical continuity that is not infrequently overlooked. Retracing the roots and de-sedimenting the origins of such an orthodoxy can help counter those recurrent attempts to present neo-liberal ideas as somehow new, previously untried, and eminently sound, thus veiling their history of failure and ideological partiality. This is also important in relation to the fact that in mainstream development discourse the role of the market, the place of free trade and investment, the functioning of private enterprise and the logic of capital accumulation are pervasively screened from any probing political critique, so that while the state and the public sphere have been continually monitored, assessed and subjected to far-reaching critique, there has been less space allocated for a critical rethinking of the role of the private sector in society.

Second, when one considers the critiques of 'structural adjustment', of monetarist policy and of neo-liberal strategy in general, it is worthwhile remembering that the framing of the economic also insinuates a conception of the individual and the relation of the individual to society.[10] The anchoring of the individual to the bedrock of possession, ownership and acquisitiveness gives meaning to the notion of possessive individualism, and has to be distinguished from the richer and potentially more open term of individuality, which connects to the complex issues of identity and difference, taken up in chapter 6.

Last, neo-liberalism is embedded in a political philosophy, which is not always made explicit, but which aspires to be universal and continuous, and the belief in the universalization of Western liberal democracy as the final form of human government represents one example of such an overarching doctrine (Fukuyama 1992). I shall return to this aspect of the politics of the universal below.

Exploring the Conceptual Dynamics of Neo-liberal Thinking

As was noted at the outset of the chapter, a key element of the neo-liberal enframing of development thought and practice was initially expressed through the organizing concept of structural adjustment. In 1981, for example, the World Bank reported that in fiscal 1980 the Executive Directors approved the initiation by the Bank of 'structural adjustment lending'.[11] This kind of lending was designed to encourage major changes in the policies of developing countries so as reduce their current-account deficits to more manageable proportions (World Bank 1981b: 69). Although couched in an ostensibly neutral and technical language, the 1981 Annual Report did signal the importance of policy and institutional changes that were to be linked to structural adjustment lending operations. Moreover, the controversial nature of structural adjustment was already alluded to, with the statement that 'the difficulty that governments find in gaining *political* acceptance for the adoption and implementation of structural-adjustment programs has been and continues to be the single most important obstacle to rapid progress by the Bank with structural adjustment assistance' (World Bank 1981b: 70, emphasis added).

The implementation of structural adjustment lending was legitimized in the context of the mounting debt crisis affecting the countries of the South. For example, in the wake of the recycling of petro-dollars and the rapid growth in both direct and indirect investments, the foreign debt of Third World countries increased from US$67.7 billion in 1970 to $438.7 billion by the end of 1980 (World Bank 1981a: 57), a figure that rose to more than $700 billion by 1985, representing as much as 33 per cent of the developing countries' GNP (World Bank 1986: 32). By 1997, this figure had increased slightly to just under 35 per cent (*The Economist*, 27 Feb. 1999: 124). In the 1970s and 1980s, much of the expanding foreign debt was concentrated in the so-called 'middle income countries', such as Mexico, Brazil, Egypt, Thailand, the Philippines and South Korea.

In Latin America, for instance, the external public debt as a proportion of GNP not only increased in the more industrialized countries of Brazil, Mexico, Argentina and Venezuela, but also in other smaller economies the increases were quite dramatic, as exemplified by the fact that from 1970 to 1984 debt as a percentage of GNP increased from 11.3 to 51.9 in Uruguay, from 12.6 to 59.4 in Peru, from 11.7 to 73.1 in Ecuador and from 13.8 to 104.2 in Costa Rica (World Bank 1986: 214–15). As the problems associated with mounting Third World debt intensified, with the Mexican debt crisis of 1982 being taken as a pivotal event, structural adjustment lending through the coordinated efforts of the IMF and the World Bank developed a crucial momentum. It led, as is well-known (and as I mentioned earlier in relation to the Bolivian case), to a package of measures which included, *inter alia*, the increased opening up of the economies of the South to the world market, the privatization of state enterprises, devaluation of national currencies, and new forms of deregulation of the economy to encourage export production. These policy changes which were initiated in the 1980s were carried through into the 1990s (see note 4).

The neo-liberal doctrine of development thus began with the formulation and deployment of structural adjustment policies which constituted a new form of intervention *in* and surveillance *over* economies of

the South. The social and economic effects of these policies and programmes have been diagnosed in detail elsewhere,[12] and what I want to
concentrate on here is the question of the changing conceptual and
thematic orientation of neo-liberalism. I want to identify the shifting
conceptual emphases and attempt to locate these moving sites of meaning geopolitically. Far from being static, the neo-liberal discourse on
development and change can be more usefully approached as being
mobile and multi-dimensional, even though it is also organized around
the founding signifiers of profitability, competition and individualism.

My argument is that the neo-liberal perspective on development from
the early 1980s to the present has been characterized by three phases.
The first phase can be denoted by 'structural adjustment', the second by
'good governance' and the third by 'social capital'. The second phase
began around the end of the 1980s, being importantly linked with the fall
of the Berlin Wall and the geopolitical watershed of 1989, and the social
capital phase emerged around the middle of the 1990s. I want to emphasize that these 'phases' or changing modalities of neo-liberal thought
overlap: that is, good governance does not replace structural adjustment
and social capital does not replace or supersede good governance. Rather
these shifts represent an extension of the discursive terrain, a colonization of other domains of knowledge so that by the beginning of the
twenty-first century, the economy, the state and civil society have been
represented and situated as part of an evolving regime of truth.

What is at stake here is the continual *mobilization* of concepts for the
diagnosis and treatment of development problems. This dynamic reflects
the impact of a rapidly changing world and the perceived need and
responsibility to respond to that reality. Always a crucial ingredient of
any hegemonic project is the will and capacity to provide intellectual
leadership, to create and disseminate the ideas, notions and concepts that
will provide a regime of truth. Specifically, a regular series of statements,
emphases, themes, priorities and recommendations need to be created as
a way of providing a vision, an explanation and also a frame for the
practical construction of a strategy that will be deployed through the
appropriate international and national organizations and apparatuses of
rule. In this way a particular discourse becomes *effective* in the 'real
world'. For the success of such a project, there has to be a continuing
dynamic and also a certain kind of effective 'out-reach' in the sense of
discursive coverage or comprehensiveness. To focus only on the 'economy' and to leave out of account an explicitly structured position on the
changing relations with state powers and civil society developments

would be to limit and undermine that aimed for effectiveness and practicality. The broader the compass of conceptual intervention the wider the ambit of power effects. Let me illustrate this idea by first of all referring to an important passage from the OECD's 1992 Development Cooperation Report.

In setting out a position on the role of development assistance and aid agencies, the OECD made the following observation:

> The key to mastering development and other global challenges is the building of fundamental institutional and economic capacities in the developing countries'.

In this context, it was stated that capacity building in the developing world was to be focussed on three central and interdependent elements: effective and competitive market systems; effective and accountable government systems and the broad-based enhancement of human capital and participation in economic, social and political life (OECD 1992: 49). These three elements, which I have defined in relation to 'structural adjustment', 'good governance' and 'social capital', were similarly highlighted in a paper by Landell-Mills (1992), former Senior Policy Adviser for the World Bank's Africa Region. Landell-Mills gave considerable significance to the role of NGOs in development performance and also to the potentially positive place of grassroots organizations in the reform of civil societies in Africa. The importance of NGOs for the World Bank was illustrated by the fact that in 1991 as many as 44 of the World Bank's assisted projects in Africa were designed and implemented in partnership with local NGOs as compared to a figure of 7 per year for the 1973–87 period (Landell-Mills 1992: 565). The conclusion from this analysis was that there are close links between governance, cultural relevance and the components of civil society. The components of civil society, such as the NGOs and grassroots organizations, are then envisaged as being strengthened by the 'economic liberalization and privatization measures that typically form a key part of the on-going structural adjustments being undertaken in most countries' (1992: 567).

The emergence by the early 1990s of these three 'central and interdependent elements' of official development discourse can be seen as an attempt to provide a comprehensive vision of the conceptual and practical terrain – a vision which has been developed through the 1990s. Before examining the combination of these elements and the way they have emerged as part of an ensemble of meanings and practices, it is important to consider the first shift or extension of the conceptual terrain which relates to the introduction in the late 1980s and early 1990s of the term 'good governance'.

In the World Bank's Annual Report for 1992, it was noted that by 1991 typically two out of every three Bank operations included components that explicitly supported private-sector development compared to a figure of about 40 per cent for 1988. The World Bank identified three tasks that were seen as essential to the promotion of private-sector development:

- the creation of an affirmative business environment;
- the restructuring of the public sector;
- the development of the financial sector for entrepreneurial activities.

It was strongly argued that public sector restructuring involved both improving efficiency in the critical functions of the state, such as the provision of social and physical infrastructure, and creating space for private initiative through a *'shift in the boundary between the public and private sectors'* (World Bank 1992a: 61, emphasis added). This shifting of the boundary was to be crucially aided by the role played by privatization, which was seen as a process that would open up new opportunities for private investors and free government resources and administrative skills for high-priority activities. The privatization strategy was to be carried out through the 'orderly sale of state-owned enterprises'. It was further stipulated that once government officials had been freed from the fiscal and administrative drain caused by public-enterprise problems, they could concentrate on those activities that are 'best suited to the public sector: the provision of essential public goods and infrastructure in support of development and the funding of programs in the social sectors that both assist the poor and help develop human-capital resources' (World Bank 1992a: 147).

This clear specification of the functions and objectives of the state was further elaborated in the same year in the World Bank's (1992c) paper on *Governance and Development*. Governance was defined as the 'manner in which power is exercised in the management of a country's economic and social resources for development', and 'good governance' was specified in terms of 'sound development management', with the further stipulation that 'good governance is central to creating and sustaining an environment which fosters strong and equitable development, and it is an essential complement to sound economic policies' (1992c: 1).

How do we interpret this shift of emphasis towards public-sector reforms and a new conceptualization of the importance of governance in economic development? The World Bank's own explanation relates to two factors. First, in the context of a Bank study of development problems in sub-Saharan Africa, published in 1989, a 'crisis of governance' was identified in relation to the failure of its structural adjustment programmes. Second, the rapid geopolitical changes in Eastern Europe, Latin America and other parts of the South, changes that were brought in train by the collapse of the Berlin Wall in 1989 placed centre-stage the need to rethink the role of the state in social and economic development. In this context, the World Bank (1992c: 5) noted that a changed role for the state was to be reflected in four areas:

- the creation of an enabling environment for development;
- larger responsibilities for the private sector;
- a reduction in direct government involvement in production and commercial activity; and
- the devolution of power from the centre to lower levels of government.

The perceived failure of structural adjustment policies in sub-Saharan Africa, taken together with the key importance of the 1989 geopolitical watershed in world affairs, stimulated a more explicit consideration of the state and the problems of governance. Schatz (1996: 241) appropriately commented that basic to the World Bank approach was the problematic belief that if the market was liberated from the incubus of government, it would achieve at least a minimally acceptable rate of development, and would 'moreover, improve as time allows the salutary competitive processes to work their effects'. Schatz's critical observation was shared by a number of authors – (see e.g. Moore 1996 and Schmitz 1995).

Similarly, Leftwich (1994: 368), in his appraisal of the World Bank's governance strategy, pointed to the fact that although the Bank committed itself to the ostensibly apolitical and largely technical strategy of improving governance, the 'apparently politically-neutral recommendations presupposed profound political change and represented not simply an economic vision but also a political one', since the kind of state that was being advocated had as its central role the encouragement of the play of market forces. A similar observation was made by Hewitt de Alcántara (1998: 107), who noted that the term 'governance' enabled multilateral banks and agencies to couch sensitive political questions in essentially technical terms. Lastly, for George and Sabelli (1994: 142) not only was the World Bank's governance 'an instrument of control' and an additional conditionality, but it offered a further opportunity to 'instil Western political values in borrowing countries and to fault them if things go wrong'.

The translation of political questions into what appear to be technical issues, or themes of public-sector management, accountability, a legal framework for development and information and transparency, can be interpreted as a politics of depoliticization, or, as suggested in previous chapters, as an example of the 'pacification of the political' – that is, placing a set of contestable orientations under the control of a settled system of understandings and priorities. This analytical suggestion can be illustrated in relation to the World Bank's proposed actions on governance for which six options, depending upon country circumstances, were itemized:

a) to assist with governmental reform (e.g. with reference to a legal and procurement framework or decentralization programmes);
b) to persuade governments through dialogue of the need for reforms (e.g. in relation to public sector management assessments and financial accountability and legal sector reviews);
c) to craft country lending strategies and levels to take account of the effect of governance on development performance;
d) to help countries deal with the especially complex issues of poverty and the environment;
e) to project a long-term vision of an 'enabling environment' for the private sector;
f) to improve implementation performance through greater efforts to assist borrowers in 'building ownership of adjustment programs', to foster an 'implementation culture' within the Bank and to promote greater understanding of the social and political structure of the countries where policies are being implemented (see World Bank 1992c: 52–3).

From these specifications there is clear evidence of a new kind of interventionist strategy which, taken together with structural adjustment policies, represented both an extension and an intensification of a powerful apparatus of rule. Within an evolving rationality of persuasion and penetration, the economy and the state were being integrated into a broader discursive frame. Particularly significant in terms of the project of discursive persuasion, was the intention to assist borrowers in 'building ownership of adjustment programs'. In other words, as in any effective project of hegemony, the subjects being 'hailed' or persuaded and incorporated into the project need to be convinced that it is their project too, that they 'own' the ideas and the practices as much as the originator of the project. The persuasion, if it is to be successful, needs to have a certain kind of plasticity whereby the receivers can also feel that they have the possibility to mould the project themselves in ways which seem pertinent to their own perceived objectives. This is also to be noted in the way the World Bank, already in 1992, gave significance to popular participation in new projects of governance, commenting, for example, that governance is a 'plant that needs constant tending', whereby participating citizens 'need to demand good governance' (World Bank 1992c: 11).

This kind of persuasion or advocacy is repeated in the OECD Development Cooperation Report for 1994, where in a general developmental context it is recommended that recipient governments should be encouraged to establish a 'culture' that facilitates the growth of the private sector. At the same time, it is argued that the reform process, especially in countries threatened by 'adjustment fatigue', should be bolstered by 'helping governments educate their publics about benefits that have accrued from ongoing reforms and expected future benefits' (OECD 1994: 15).

A symptomatic and problematic feature of the desire to 'build ownership' concerns the issue of West/non-West relations. Above, reference was made to George and Sabelli's critical point concerning the inculcation of Western values in the matter of governance, and it is important here to recall that the World Bank in its report on governance and development did signal the relevance of this issue. For example, it was argued that the debate on governance needs to take into account cultural differences. This suggestion was illustrated in relation to the theme of contractual agreements and the variability across cultures of conceptions of rights and obligations, including, for instance, the role of binding ethnic or kinship ties. Furthermore, it was noted that the 'spread of political and legal systems modelled on Western traditions may lead to the simultaneous existence of two sets of norms and institutions for dealing with ... rights and obligations, with Western notions of the rule of law, private property rights and contracts superimposed on ideas such as "consensus", "communal property" and "reciprocity", which have evolved over long periods of time in many non-Western cultural settings' (World Bank 1992c: 8). The Bank went on to remark that even though these varied ways of anchoring social rights and obligations 'may or may not hamper the functioning of modern economic institutions, such as the market, it is important to be aware of them when pursuing policy reforms and institution building' (ibid.).

This interesting and important passage did not, however, act as a basis for subsequent elaboration. Instead, the potential relevance of cultural sensitivity and also the problems involved in superimposing a Western model, both in respect of the definition of the role, function and understanding of the market, as well as the approach to social and political institutions and differential cultural values, were left unexplored.

While it is necessary to highlight the way one conceptualization of governance was formulated in relation to the already existing strategy of structural adjustment, and to underline the extension of the terrain of intervention to include matters of public-sector management, it is also important to consider the new significance being given to the spatiality of governmental structures. This was reflected in the call for decentralization, which again revealed the tension between notions of efficiency and participation.

A significant place for governmental decentralization first surfaced in World Bank development strategy in the World Bank's 1988 *World Development Report*, where the financing of local government was connected to 'fiscal decentralization'. This report, in a discussion of state-owned enterprises, offered the view that because the barriers to 'full and rapid privatization are often daunting, intermediate solutions...are often more feasible'. These intermediate solutions were then identified as subcontracting, leasing or the introduction of private competition and 'fiscal decentralization' (World Bank 1988: 10).

The idea of decentralization has always had a certain charm. It can be deployed as a means of breaking down the solid blocks of a central bureaucratic structure, or invoked as a primary step towards a more sustainable pattern of social and economic development, or linked to calls for more participation in the decision-making process as a whole. Decentralization, in the case of the World Bank and other institutions of international development, has also been used in conjunction with privatization, deregulation and the rolling back of many of the economic and social functions of the state. In the early 1980s, an expert on decentralization remarked that as larger numbers of small-scale projects and area-wide, multi-sector, 'integrated' projects increase in importance, their overall objective, which is to 'reach the rural poor', is made more problematic by the continued existence of over-centralized management.

Faced by the problem of over-centralization, it was argued that if the alleviation of poverty and balanced development were to be achieved, then popular participation and the decentralization of authority ought to be encouraged (Rondinelli 1981: 133).

In the World Bank paper on governance and development, a link is made between decentralization and both efficiency and participation. Thus, it is posited that decentralization can lead to key improvements in efficiency by 'reducing overloading of central government functions', while also 'improving access to decision-making and participation at lower levels of government' (World Bank 1992c: 21). This combined emphasis on effectiveness and participation is important to keep in mind because, while the first is to be closely tied into privatization and public-sector reforms, the second is crucial to the legitimacy of the changes being engineered. This unstable combination is captured in a World Bank definition of 'successful decentralization', which is seen as a means for the improvement of the 'efficiency and responsiveness of the public sector' while at the same time 'accommodating potentially explosive political forces' (World Bank 2000a: 107).[13] Of course, the actual balance of emphasis has tended to vary, although it would seem that overall the connection with fiscal rectitude and private-sector development has received more attention.

This is reflected in an Inter-American Development Bank special report on fiscal decentralization in Latin America, where the stress falls on efficiency, as exemplified in the statement that 'the search for more efficient resource allocation is the leitmotiv of public sector decentralization' (IDB 1994: 175). Similarly, at a joint international forum on democracy, decentralization and deficits in Latin America, organized by the OECD and the IDB, the following points were recorded: (a) that democratic institutions were becoming more decentralized, as reflected by the fact that while only three countries in 1980 had direct elections for mayor, by 1998, 17 did, and (b) the democratic/decentralizing trend was consistent with improved fiscal discipline, so that while only 4 countries in 1983 had deficits below 3 per cent of GDP, by 1998 16 countries did (see Hausmann 1998: 13).

More recently, the World Bank, in its Development Report on the state, singled out decentralization as bringing many benefits in much of Latin America, China, India and many other parts of the world. For the Bank, decentralization could improve the 'quality of government' and the 'representation of local business and citizens' interests'. More specifically, the Report contended that '*competition* among provinces, cities and localities can spur the development of more-effective policies and programs' (World Bank 1997: 11, emphasis added). This is a key assertion of neo-liberal strategy – the notion that competition, rather than collaboration and cooperation, will lead to further growth and development; but how this is to be squared with the importance given to participation remains an open question. Its further examination leads quite appropriately into the third interconnected element of development strategy, set out in the OECD 1992 Development Report, and referred to above – namely the place of 'human capital' and participation, an element I shall briefly discuss under the rubric of 'social capital'.

As noted previously, when analysing the dynamic of neo-liberal discourse, the extension of the conceptual terrain has to be situated in relation to the perception of change, both as regards the actual effects of official policies, e.g. around the impact of structural adjustment, but also in terms of the perception of socio-economic and political change as a whole. To illustrate this suggestion, it is useful to refer to an address given by James Wolfensohn, President of the World Bank Group, to its Board of Governors in 1998 – the title of the address was *The Other Crisis*, and the theme was the 'human dimension' of the global financial crisis.

In outlining what Wolfensohn referred to as a 'new approach', certain potential dangers on the road to successful development were identified. For example, although countries could move toward a market economy, privatize, break up state monopolies, and reduce state subsidies, if they did not introduce safety nets and if there was no social and political consensus for reform, then their development could be endangered. Equally, while some countries may attract private capital, deliver growth and invest in some of their people, if they also marginalize the poor, women and indigenous minorities and fail to introduce a policy of inclusion, then also their development will be undermined. Consequently, Wolfensohn concludes, in the global economy of today, it is the '*totality* of change in a country that matters' (Wolfensohn 1998: 11, emphasis in original). This means that development is not just about adjustment, sound budgets, education and health and technocratic fixes – it is about putting *all* (emphasis in the original) the component parts in place – together and in harmony. Development involves a totality of effort. 'Too often,' Wolfensohn concedes, 'we have focused too much on the economics, without a sufficient understanding of the social, the political, the environmental, and the cultural aspects of society' (Wolfensohn 1998: 12).

The idea of a totality of change is envisaged in terms of two mutually sustaining frameworks. First, the IMF framework which is needed to evaluate macro-economic performance for all countries, and second, a framework which deals with the progress in structural reforms necessary for long-term growth. This second framework is referred to as a development framework which would include (a) the essentials of good governance, (b) the regulatory and institutional fundamentals basic to a market economy, (c) the promotion of social inclusion, especially through education for all, especially women and girls, (d) the improvement of public services and infrastructure, in both urban and rural areas, and (e) the guaranteeing of environmental and human sustainability wherein it needs to be ensured that the culture of every country is nurtured and enriched. These five interrelated elements are then linked to the significance of 'ownership' (whereby governments must be in the driver's seat), the importance of participation, and finally to the need for partnership, in which the donor community and the Bank are appropriately seen as the catalysts rather than the cartographers of change.

This 'second framework' received further emphasis in the World Bank's 1999 Annual Report, where the concept of a 'Comprehensive Development Framework' captured the key points. The central features of this framework were defined in terms of contributing more effectively to the human, social and structural aspects of development, of strengthening partnership and enhancing inclusion and promoting participation by civil society, communities, local governments and NGO's (World Bank 2000b: 26).

In this same Annual Report, the emphasis being given to social development is linked into the significance of what is referred to as 'social capital'. Here the World Bank broadly defines social capital as the 'institutions, relationships, attitudes and values that govern interactions among people in society and contribute to economic and social development' (World Bank 2000b: 122). In the accompanying World Development Report for 2000/1, social capital is again profiled in the context of building social institutions, which are actually defined as different dimensions of social capital. These dimensions are seen as follows: bonding social capital relates to strong ties connecting family members, neighbours, close friends or business associates; bridging social capital implies weaker horizontal ties connecting individuals from different ethnic and occupational backgrounds; and linking social capital consists of the vertical ties between poor people and people in positions of influence in formal organizations such as banks or agricultural extension offices (World Bank 2000a: 128–9). In the World Development Report for 2003, a further distinction is introduced whereby 'traditional social capital' is defined in relation to trust and sharing, which if combined with 'modern assets' such as educated men and women can help communities move in the direction of development (World Bank 2003: 75).

As can be appreciated from the World Bank definition of 'social capital', a very wide spectrum of social and political phenomena are encapsulated into one rather econocentric concept. Trust, association, solidarity, cooperation, networks, movements and so on are amalgamated under one limiting label. The stress on 'capital', albeit social capital, is linked to the earlier concept of 'human capital',[14] and is reflective of a reductionist tendency in economics and economic sociology where narrowly conceived categories such as calculation, economic rationality and individualized motivation and achievement are given a much broader explanatory significance whereby complex social tendencies are strait-jacketed into confining concepts. As one example, it might be noted that the World Bank, in connecting social capital to development, suggests that social networks and organizations are 'clearly key assets in the portfolio of resources drawn on by poor people to manage risk and opportunity' (World Bank 2000a: 129). In this context, the World Bank is closely following the approach used by Putnam (1993) in his work on democracy in Italy, where social capital is likened to a 'stock' of norms of reciprocity and networks of civic engagement – social capital, for Putnam, refers to features of social organization, such as trust, norms and networks that can improve the posited efficiency of society.[15]

Further critical assessment of the social capital category and its use in development discourse would certainly be possible (for an extended analysis see Fine 2001), but rather than pursue such a pathway it is now necessary to connect together the three 'phases' of the neo-liberal discourse on development.

Ideas and Effects

I have argued in the above section that the neo-liberal discourse of development as constructed by international organizations such as the World Bank and the OECD has a dynamic that encompasses economy (structural adjustment), state (good governance) and civil society (social capital), and also includes a vision of the individual as being market-oriented. It is therefore a discourse with totalizing ambitions, as Wolfen-sohn intimated in his 1998 paper. At the same time its cohesiveness rests on certain central or master signifiers. Competition, privatization, market-led development, efficiency, and sound management provide examples of the economic anchor, but notions of good governance, fiscal decentralization and a minimal or (by the late 1990s) effective state (World Bank 1997) are important as complementary signifiers of the role of politics in a market-led development process. What we have seen evolving since the early 1980s is a discourse that while being based in a certain view of the economy, has been extended out to colonize the social and political terrain to give it a broader coverage and greater potential explanatory power. Also, in terms of influence and leadership, a discourse which, while being based in key ideas on the economy, moves out to construct a wider interpretation of the social and political dimensions of development, is more likely to become hegemonic. Moreover, the will to be hegemonic has to be seen as being linked to two further factors.

First, the neo-liberal framing of development combines a regularity of conceptual and practical orientation with an institutionalized power that produces specific effects on the policies and practice of Third World governments. The World Bank and the IMF and (on trade) the WTO are not only guardians of certain organizing ideas, but they are designed to make and implement policy. At the beginning of the chapter I briefly discussed the Bolivian example in terms of the effects of structural adjustment, and similar points can be made for other economies of the South. Badie (2000: 43–5), for example, refers to the revealing case of Algeria, a country with a government that in the 1980s still gave consid-erable weight to socialist ideas, and denounced the World Bank as an imperialist institution. After a World Bank visit in 1989, the state was required to withdraw from the 'economic machinery', leaving it to other agents, enterprises and households. The private sector was recognized as crucial in development and state farms were to be terminated. Many similar examples could be cited. The overall point is that neo-liberal hegemony has combined the power of discursive enframing with concrete effects on the strategy of Third World governments.

Second, as emphasized in many development documents through the 1990s, partnership and the local ownership of policies designed by the international institutions were seen as a key component in the success of the new development strategy. In other words, development policies on adjustment, good governance and market-led change would only be effective if adopted and made their own by Third World governments and societies – in the words of the World Bank President, we are the catalysts not the cartographers. Clearly, external projection of policy on its own would not be enough. It is in this kind of context that the Mexican sociologist Zermeño (1996: 104–14) has argued that there is a 'dependent neo-liberalism' whereby governments and elites in the South not only accept the donor strategy but embrace it with vigour and enterprise. It is here, however, that we need to draw an important distinction.

In terms of political effects, the three main components of the neo-liberal perspective on development have had a differentiated impact. Structural adjustment has gone together with clearly identifiable effects on government policy, whereas good governance has had far fewer concrete policy effects, as a range of authors have shown (see e.g. Gibbon 1993 and Doornbos 2001). Social capital has, so far, had little tangible policy effect, although evidently its power is more appropriately understood as being complementary to that of the central categories of privatization, commodification, competition and open markets. What I am arguing here therefore is that in terms of practical effect, structural adjustment has been more incisive in its influence. However, this point needs to be further contextualized by introducing another, more explicitly geopolitical factor into the discussion.

Market-based Democracy and Geopolitical Intervention

While it has been the central aim of this chapter to consider the emergence, roots and evolution of neo-liberal ideas on development and North–South relations, such an appraisal would be incomplete without at least a brief analytical connection being made to the changing politics of the period under review, and especially the 1980s. This cannot be done in detail, but something must be said concerning the Reagan Doctrine and the re-assertion of Cold War politics until 1989. The Reagan Doctrine is important as a precursor for key aspects of Washington strategy post 9/11, a topic to which I shall return in chapter 7, but it is also relevant to one of our chief themes, the geopolitics of power and representation.

While neo-liberalism was diffused by the Washington administration during the 1980s, it was closely associated with market-based notions of democracy, conceived in a US mould. In 1982 President Reagan, in a speech before the British parliament, announced that the US would pursue a new programme to promote democracy around the world. It was called 'Project for Democracy' and institutionalized as the National Endowment for Democracy (NED), which was funded by the US Congress. Between 1984 and 1992, the NED and other branches of the US state mounted 'democracy promotion' programs in 109 countries around the world, including 30 countries in Africa, 24 in Asia and 26 in Latin America and the Caribbean. Democracy promotion, as it was termed, included leadership training, education, strengthening the institutions of democracy, conveying ideas and information and developing institutional and personal ties (Robinson 1996: 332, 107). The aim was not only to promote a particular view of democracy, founded in support for 'free and fair elections', but also, as part of the Reagan Doctrine, it was designed to confront what was seen as the expanding threat of communism.

The Reagan Doctrine represented a reduced role for containment and the beginnings of a more aggressive strategy of attempting to 'roll back' communism. It was defined in 1985 by conservative columnist Charles Krauthammer as proclaiming 'overt and unabashed American support for anti-communist revolutions', with its grounds being in 'justice, necessity and democratic tradition' (qtd in Scott 1996: 1; see also Kenworthy 1995: 70–1). The promotion of what has been called 'low-intensity democracy' (Robinson 1996) was intertwined with a will to intervene in new ways to confront Third World governments such as the Sandinista government in Nicaragua which had broken away from dependence on the US. Although the Sandinistas organized elections in 1984, elections which were considered to be fair and free by independent observers, (see, for example, Cornelius 1986 and Robinson & Norsworthy 1985), the Reagan administration continued to finance the Contra guerrillas and the subversion of a democratically elected government (Cottam 1994: 130). In other words, what was at issue here was not only the definition of democracy but the Reagan administration's unwillingness to accept Third World governmental independence from US power, especially in its own Central American 'backyard'.

Washington's hostility towards the Sandinista government was clearly reflected in the financing of the Contra guerrillas[16] and acts of economic sabotage which forced the Sandinista government to allocate an increasing amount of the national budget to defence – an estimated 60 per cent by 1985. But also, as Robinson (1996: 238) shows, towards the end of the 1980s the Reagan administration successfully pressurized the World Bank to block international assistance to Nicaragua, reinforcing the isolation and economic dependence of the Sandinista government.

Washington's overall strategy of undermining and terminating the Sandi-nista's project came to successful fruition in the 1990 elections with the defeat of the Sandinista party and the coming to power of a pro-US government led by Violeta Chamorro. Following the election, the dismantling of Sandinista institutions involved a radical restructuring of the state, with the introduction of sweeping economic policy reforms, reforms in the police, the constitution, social policy, the re-assertion of a predominant role for private capital and the opening up of the economy through market liberalization. With the new government being depend-ent upon US support, the Nicaraguan state was transformed into a neo-liberal or market state, and came in the 1990s to reflect the broad trend of neo-liberal globalization with its streamlining of the state and prioritization of market-led development.

Washington-based policy both in the 1990s and beyond reflected and supported the World Bank and IMF strategy of structural adjustment and monetarist economics. The emphasis on market-based democracy, with 'free and fair elections' and the Schumpeterian notion of a competitive struggle for the people's vote[17] acting as guiding tenets, was also allied to the World Bank's notion of good governance. In a sense the minimal or effective state parallels what can be considered as a notion of minimal democracy, where the role of political institutions and politics in general is to support the primary engine of market-led development. The extra specificity of US foreign policy was that a market-based vision of dem-ocracy was part of the Cold War politics of the 1980s, with an overall strategy of rolling back what was seen as the menace of communism in general and Marxist regimes in particular.[18] Such a strategy was reflected in the destabilization of radical governments as in the Nicaraguan case, or in the buttressing of conservative governments, such as in El Salvador and Guatemala, that were being threatened by communist-inspired guerrilla movements (Cottam 1994). Thus while the Reagan Doctrine had a neo-liberal basis it also had a geopolitical edge that was recovered and re-moulded in the post-9/11 era.

A Concluding Comment

In this chapter I have looked at the rise of neo-liberal ideas on develop-ment and change and emphasized their historical lineage, conceptual dynamics and practical impact on the economies and societies of the South. In contrast to modernization theory, neo-liberalism has given greater emphasis to private capital, to competition, accumulation, deregulation, open economies, leaner states and market-oriented progress. However, in terms of their commonality, both perspectives have provided a legitimation for the projection of Western power, based on the presumption of occidental supremacy and a belief in the benign diffusion of such power. There is also a parallel in the way that, while modernization theory came to give more weight to order and political control, similarly the neo-liberal perspective mutated from structural adjustment to a more overtly political focus on the nature of government and social organization. While neo-liberal ideas have acquired great power to the point of being globally hegemonic, there has also emerged increasing opposition to their influence, as we shall see in chapter 8, when the theme of social movements and alternative globalization is explored. What is now needed is an analysis of the counter-currents to both modernization and neo-liberalism. This will be the subject of the next two chapters.

Part III

Archipelagos of Critical Thinking

5

Societies of Insurgent Theory: The *Dependentistas* Write Back

'The social and economic transformations that alter the internal and external aspects of underdeveloped and dependent societies are actually political processes that, in present historical conditions, do not always favor national development.'

– F. H. Cardoso and E. Faletto (1979: 28)

'Latin America is the region of open veins. Everything ... has always been transmuted into European – or later – United States – capital, and as such has accumulated in distant centers of power. ... To each area has been assigned a function, always for the benefit of the foreign metropolis of the moment ...'

– E. Galeano (1973: 12)

Introduction

One of the main objectives of the last chapter was to trace the genealogy of neo-liberal thought in order to question its purported newness and to point to its previous as well as contemporary partiality. While neo-liberal ideas are often presented as if they were new and liberating, in contrast dependency perspectives are frequently portrayed as being irrelevant to the present, being part of a failed world-view. A colonizing vision of history, aptly described by Fanon (1969: 169) as seeking to distort and disfigure the past of an oppressed people, would relegate the oppositional analysis of dependency to an historical backwater.[1] What is at stake here is the function and geopolitics of memory. Memory is particularly important with reference to dependency studies, and its corollary, 'radical underdevelopment theory', since despite a current view that such older perspectives are irredeemably obsolete, there is still a sense in which many of the ideas raised by the dependency writers of the 1960s and early 1970s continue to retain a contemporary validity.

On the broad terrain of ideas and history, it was Nietzsche who suggestively argued that we need history for the sake of life, and specifically a critical history that will encourage the capacity to develop independently, and to incorporate and above all transform what is past and different.[2] The enabling feature of this part of Nietzsche's thought is that looking back can be helpful if we are doing so to make interventions in the present. But more specifically, what is relevant here is the recovery of one kind of 'disqualified', subordinated knowledge, a kind of knowledge that is rooted in the memory of oppositional encounters, or what Foucault (1980a: 83), in his discussion of genealogy and knowledge, termed the 'historical knowledge of struggles'. This emphasis on the importance of the recovery of oppositional knowledges can be applied to dependency perspectives, where a critical rethinking and reproblematization can help sustain oppositional perspectives to contemporary neo-liberalism. This is not to suggest that I want to 'rehabilitate' the *dependencia* discourse as if it were a homogeneous entity. It had its own variants and dissonances, and it was also characterized by several deficiencies that will be examined below. Therefore, my purpose is not to eulogize a past vision, but rather to ask the following questions: (a) what were the main arguments of the dependency perspective?; (b) how can the dependency vision be located geopolitically?; (c) what can be learned from the dependency viewpoint, and how relevant is it in today's world of growing inequalities and power disparities?; and (d) how can dependency be related to contemporary post-colonial analysis ?

The Challenge from the Periphery: The Emergence of a Counter-analysis

It is not my intention here to provide a comprehensive treatment of dependency thinking, since there already exist a number of valuable studies of the main currents associated with dependency and radical underdevelopment theory (see e.g. Blomström & Hettne 1984, Kay 1989, Larrain 1989, and Rist 1997). Rather my purpose in this section of the chapter is to outline some of the main arguments that emerged from the early 1960s onwards in the Third World, and especially in Latin America – arguments that were aimed at formulating a different explanation of development and underdevelopment across the North–South divide. My focus will fall not just on the more socio-economic expressions of these theoretical interventions, but also on the underlying questioning of power and politics that informed and drove much of the thinking and representation that came out of the Latin South. My suggestion would be that the dependency perspectives that emanated and spread from Latin America to other parts of the periphery were part of a vital project of *counter*-representation – that is, that together with related writings from Asian and African scholars a quite distinctive mode of interpreting West/non-West, First World/Third World or centre–periphery relations began to take shape in the 1960s and 1970s, and that this new mode of thinking provided a key challenge to the tenets of modernization theory.

If one begins with the situation in Latin America, then the first point to remember as regards the evolution of alternative theories of development is that many important facets of dependency perspectives, especially in the social and economic realm, originated with the work of the United Nations Economic Commission for Latin America (ECLA), based in Santiago de Chile and established in 1947. As is well known, one of the essential elements of ECLA's originality lay in the thesis that the formation of development and underdevelopment was a single and combined process. It was argued that the disparities between centre and periphery were reproduced through international trade, and that the colonial origins of the incorporation of Latin American economies into the capitalist system had led to the creation of an outward-oriented development model, based on the export of primary products. The argument that the terms of trade between centre and periphery were disadvantageous to the latter represented a clear challenge to the conventional theory of comparative advantage through international trade, as well as providing a firm foundation for the ECLA proposal to establish an inward-oriented development strategy.[3]

As early as 1949, the Argentinian economist and driving force of ECLA, Raúl Prebisch, observed that incomes grew faster in the centre than in the periphery, and that this widening gap had to be explained by reference to the prevailing international division of production and trade which tended to confine the periphery to the production of primary commodities. Furthermore, the socio-economic disadvantages accompanying a small or absent capital-goods sector, with all the implications of imported technology for employment, income generation and further financial dependence, were also signalled in the 1950s. These kinds of ideas, originally introduced by the ECLA school,[4] were to resurface in later elaborations of dependency studies in Latin America. Although the ECLA school was primarily concerned with questions of economic policy, economic development was very much connected to social issues, and in the 1960s the emerging problems of import-substituting industrialization were connected to the negative impact of skewed distributions of income, insufficient employment creation and the absence of land reform, features that are still prevalent across the social and economic landscapes of contemporary Latin America (see e.g. IDB 1998).

In a more radical version, a number of *dependentistas* (dependency theorists) focused on the limitations of Latin American industrialization by emphasizing the fact that industrialization was determined by foreign investment, which was based on multinational firms whose power continued to be located in the metropolitan economies of the capitalist system. The operation of these firms did not improve the distribution of income but rather led to the further concentration of wealth and power in large groups of businesses. Further, since dependent industrialization was based on an imported labour-saving technology, not enough employment was created to absorb migrant labour from the countryside, and as both rural–urban migration and population growth proceeded the bases were laid for the production of urban marginalization.

In order to develop a more systematic focus on these kinds of issues, it would seem useful to identify the following interconnected points, themes and arguments.

1. Development and external dominance

In contrast to one of the main beliefs of the modernization theorists, dependency writers took a much more critical view of the developmental role of foreign investment in the Third World. The Brazilian economist Celso Furtado (1969: 72), for example, once remarked that the US corporation seems to be an inadequate instrument for dealing with Latin American problems – the large corporation, he suggested, 'produces the same effect in an underdeveloped economy as large exotic trees introduced into an unfamiliar region: they drain all the water, dry up the land, and disturb the balance of flora and fauna.' One of Furtado's conclusions was that development policy in Latin America would only be successful if the great mass of the region's population were to be mobilized on a national basis and in accordance with national values and ideals.[5] With Furtado and many other writers within the dependency tradition, a customary connection was made between the operation of multinational corporations and the repatriation of profits, as exemplified by the Brazilian geographer de Castro (1969), who pointed to the continuing reality of profit repatriation by US firms active in Latin America in the 1950s and 1960s.

The overall approach tended to stress the limits placed on autonomous economic development by the penetration and control of foreign capital in a wide variety of economic sectors, and in contradistinction to modernization theory, the insertion of peripheral economies into the international capitalist system was predominantly envisaged as being detrimental to national development. The texture and direction of this argument did, however, vary considerably, and in the case of Cardoso and Faletto (1979), there was less emphasis placed on the negative associations of foreign capital than with other writers such as Bambirra (1974) and Marini (1975). This issue of the posited negative effects of the 'external', situated here in relation to the debate on foreign capital, represents a symptomatic dependency position. Moreover, it can be argued that such an interpretation still retains an important contemporary relevance in an era of globalization, as we shall see in chapters 7 and 8.

2. Dependence and marginality

Flowing out of the belief in the predominance of a limited, technologically and financially dependent pattern of capitalist development, it followed for a number of Latin American social scientists that 'dependent capitalism' was accompanied by an urban process of marginalization. Although marginality tended to be originally associated with traditional notions of modernization and cultural dualism,[6] by the late 1960s and early 1970s a Marxist-oriented series of interventions and debates had emerged. In general, this literature was strongly inclined to analyse the phenomenon of urban marginality in relation to the dynamic of capital accumulation and Marx's analysis of the industrial reserve army and the varying forms of relative surplus population. As a consequence, the social and political dimensions of marginalization tended to be either neglected or interpreted as epiphenomena of the economic, and the issues of social subjectivity and mobilization were interpreted in the context of class relations. While it can be justifiably suggested, as Kay (1989: 123) noted, that one of the merits of marginality analysis was the way it drew attention to the situation of a 'vast and heterogeneous mass of impoverished Third World labour' and stimulated detailed research on how the poor make a living, equally one of its defining problems was the tendency to assume that economic factors were necessarily central in social theory. And such a position has been recently reiterated by one of the original proponents of dependency thinking, Dos Santos (1998), in his retro-spective view of dependency in a Brazilian context. What is important to signal here is that although it is valuable and necessary to continue to point to the effects of a particular kind of insertion into world economy on employment structures and socio-economic participation within the Third World, at the same time we need to keep in mind the impact of broader social and political conditions.

3. Beyond the modern–traditional dichotomy

It was in the broader geopolitical context of the relations between the dominant and dependent countries that the *dependentistas* made one of their most distinctive contributions to the debate on development and social theory. They took effective aim at the notion, basic to moderniza-tion theory, that the international system could be seen in terms of a dichotomy between modern and traditional societies, and that also within the so-called developing society there was a division between a dynamic, innovative, modern sector and a backward, rural, traditional sector.

For Cardoso and Faletto (1979) there were two fundamental problems with the conventional delineation of modern and traditional societies. In the first place, the concepts of 'traditional' and 'modern' were neither broad enough to give satisfactory meaning to all existing social situ-ations, nor were they specific enough to identify the structures condition-ing the ways of life of different societies. Second, the traditional/modern split was not able to account for the linkages between different economic stages (for example, underdevelopment or development through exports or import substitution) and the various types of social structure that are designated as either 'modern' or 'traditional'. Crucially, for Cardoso and Faletto, changes in the social structure involve a series of relations among social groups, forces and classes, and within this process there are always attempts by a particular class or social force to impose its domination over society. Their approach always stressed the significance of 'historical specificities', both economic and social, and the imbrication of the national and international levels when examining development processes.

The Mexican anthropologist Stavenhagen (1968) also took issue with dualistic notions of social change, arguing that the developed areas of the underdeveloped countries operate like a pumping mechanism, drawing from their backward underdeveloped hinterland the very elements that make for their own development. Set in a context of the channelling of capital, raw materials, abundant foods and manual labour, Stavenhagen (1968: 18) contended that the backward zones of underdeveloped countries have always played the role of internal colonies in relation to the developing urban centres or the productive agricultural areas.[7]

In general, the critique of dualism emphasized the combined but uneven nature of capitalist development, and set out to highlight the complex heterogeneities of societies undergoing rapid processes of change attended by a variety of social and political conflicts. In contrast to the Western notion of a modern/traditional binary split, both international and internal to the peripheral society, Latin American social scientists such as the Peruvian sociologist Quijano (1968) drew attention to the predominance of dependency relations between developed and underdeveloped countries and to the variegated processes of social change affecting both city and countryside in ways that could not be understood within the straitjacket of a modern/traditional binary categorization.

4. Recovering historical specificity

In close association with the above criticism of the dualistic imagination, it was also noted that within modernization theory Third World societies were represented as if their histories were essentially initiated through their contact with the West. In this sense the 'traditional society' was conceived as not only lacking capital, technology, entrepreneurship, modern institutions, educational achievement, but more fundamentally as lacking a history *sui generis*. One of the successes of the Cardoso and Faletto (1979) text was the way it revealed, for Latin America, the historical existence of different types of social and economic structure. They suggested, for instance, that the expansion of capitalism in Bolivia and Venezuela, in Mexico or Peru, in Brazil and Argentina, did not have the same antecedents nor consequences, and that the differences were embedded in a variety of factors – for example, the diversity of natural resources, and the varied moments at which fractions of local classes allied or clashed with foreign interests, which in turn ushered in particular political regimes, produced distinctive ideologies, or led to alternative strategies in response to imperialist challenges in diverse moments of history.

The emphasis on historical and conceptual difference was also present in the work of African and Asian intellectuals, who although being influenced by the newly emerging ideas from Latin America, developed their own interpretations of dependence and peripheral capitalism. The Egyptian economist Samir Amin, for example, basing his Marxist-oriented approach on a series of empirical studies of West African countries, identified four main characteristics of peripheral societies: (a) the predominance of agrarian capitalism in the national sector; (b) the creation of a local, mainly merchant bourgeoisie in the wake of dominant foreign capital; (c) a tendency toward a highly specific form of bureaucratic development, which gave rise, in the work of other authors, to notions of a 'bureaucratic bourgeoisie', and (d) the incomplete and restricted nature of proletarianization (Amin 1976: 333). Similar features were discerned for Asian societies (Banaji 1973), and what was particularly being stressed, *contra* Western development theory, was that the histories of capitalism at the periphery could not be explained through the application of a universalist scheme rooted in a limited vision of Western modernization.

Finally, under this fourth heading, it needs to be added that radical scholars of the early 1970s also re-asserted the historical significance of the wealth, diversity and cultural achievements of the societies of the periphery. Walter Rodney (1972), for instance, in his examination of the ways in which Europe underdeveloped Africa, referred to the statement of a Gold Coast nationalist in the 1920s who proudly contended that before the advent of the British, 'we were a developed people, having our own institutions, and having our own ideas of government' (Rodney 1972: 40).[8] Similarly, in the West, critical scholars pointed to the cultural richness and technical achievements of non-Western societies before their integration into colonial systems (see Griffin 1969: 34ff).

5. Prioritizing sovereignty

Running through a wide range of dependency and related writings from the 1960s and 1970s, one can locate a concern with questions of national self-determination, autonomy and sovereignty. Initially, perhaps, the main emphasis fell on economic themes, and one of the later criticisms to be levelled at the *dependentistas* concerned their tendency to over-concentrate on issues of technological and financial penetration, while giving much less attention to cultural and political complexities. However, it can be suggested that in a number of texts, particular weight was given to questions of independent development and autonomy, where there was no founding prioritization of the economic. Cardoso and Faletto (1979), for instance, argued that political conditions, such as revolutionary change in countries like China and the Soviet Union, had made it possible for development and autonomy to be combined. In contrast, in the countries of Latin America, the possibilities of a national mobilization of resources and organizational creativity were much more curtailed due to the dependent nature of their social and political systems.

Expressed more directly, the Brazilian political scientist Ianni (1971) linked structural dependency with imperialism and the militarization of Latin American societies, suggesting that political power was the essential element in the reproduction of dependency relations. He gave particular attention to the impact within dependent societies of a series of subordinating relations. For example, through the adoption of US geo-political doctrines on security, Latin American countries had experienced a subordinating political-military alignment, and similarly metropolitan organizations through their control over the means of communication, television programmes, the translation and production of books and reviews, etc., had produced within the dependent society the phenomenon of a subordinating cultural alienation. Ianni also stressed the connection between what he called the 'culture of violence', clearly visible in the persistent phenomenon of military interventionism, and illegal changes of government. In this context, Ianni underscored the reality of imperialism and US hegemony in Latin America.[9]

Overall his concern reflected a belief in the need to situate dependency relations in a context formed by geopolitical power, ideological conflict, cultural domination and the struggle for national sovereignty. Ianni represented a stream of dependency thinking that stressed the political and cultural facets of US–Latin American relations and the centrality of political autonomy and sovereignty. Unfortunately, this particular vision has frequently been neglected in contemporary considerations of dependency analysis, even though it still retains a current relevance.

6. A politics of difference

Closely linked to the issues of autonomy and sovereignty, issues which have not always received the attention they deserve within the overall reception of dependency and related radical perspectives, one has the contemporary theme of the politics of identity, difference and recognition. Although the *dependentistas* have not always been associated with critiques of Western ethnocentrism (Gülalp 1998) or with debates on the politics of difference, I would argue that a significant and persistent undercurrent of all the writing that may be placed under the dependency rubric concerns the right to recognition and independent identity, whether it be couched in more specific socio-economic terms or more broadly expressed in the arena of politics and philosophy. Although not customarily located within the conceptual domain of dependency, the Mexican philosopher Leopoldo Zea has articulated a series of ideas that can be located within such a domain, and especially in relation to the dynamic of identity and difference.

In Zea's work, history and philosophy are conjoined in a project of critical analysis which foregrounds questions of dependence, difference and power. While dependence is treated within a long historical perspective that examines the impact of the colonial encounter, recalling Simón Bolívar's condemnation of the imposition of other systems of government, copied from peoples with a different history (Zea 1963 and 1989), there is also analysis of the more contemporary influences of the United States within Hispanic America. In particular, Zea problematizes the incompatibility of democratic government with US support for tyrannies in Latin America. He raises critical questions about the projection of United States power and also about the construction of an independent identity for Hispanic America when confronted by such a modern imperial power (Zea 1970). Some of his ideas express a continuity with those of the nineteenth-century Cuban revolutionary intellectual José Martí, who, as was mentioned in chapter 2, stressed the need for 'our America' to avoid becoming a new satellite of the 'colossus of the North'. In addition, Zea's ideas can be associated with the Mexican historian O'Gorman's (1972) path-breaking study of the 'invention of America', with its critical treatment of the meanings of American history across the Old World/New World divide, where the politics of identity and difference are located in the controversies of origin and discovery.

There are two points that can be made here. First, not only does dependency thinking belong to a critical intellectual tradition that has its roots in much earlier considerations of the place of a Latin South in the hemisphere of the Americas, but secondly, such thinking can be broadened and strengthened by its association with philosophical interventions that have been concerned with identity, recognition and the rights of subordinated peoples. Moreover, the work of authors such as Zea and more recently Dussel (1998) underlines the significance of the history of geopolitical encounters in the Americas. It enables us to keep in mind the fact that a primary concern of the *dependentistas* fell into a history of ideas that had been deeply influenced by the legacy of conquest and colonialism and the continuation under different forms of the geopolitics of invasiveness. Thus, for instance, embryonic representations of dependency thinking had already surfaced in the middle of the twentieth century.

For example, just after the Second World War, a coalition of Mexican civil society organizations, including both the National Chamber of Commerce and the Confederation of Mexican Workers, expressed opposition to US proposals for an 'economic charter for the Americas', the Clayton Plan. This plan was based on the notion of free trade, including the encouragement of private enterprise. At a Congress of Industrial Development held in 1947 in Mexico City, a Mexican critique of the Clayton Plan was formulated in which it was stated that the Plan was an attempt by the United States to construct world dominion, assigning itself the role of a metropolitan country, while other countries were regarded as satellite states. As Niess (1990: 139) points out, the formulation used could have been written by the later theorists of *dependencia*. Hence, it can be argued that it would be insufficient to undertake an examination of dependency ideas in Latin America without taking into full account the longevity of critical Latin American perceptions on the history of US–Latin American relations, and the place therein of the imperial power of the United States.

What these counter-conceptualizations sought to emphasize was that it was vital to contest the pervasiveness of imperial representations as an essential part of any project of authentic independence and sovereignty. This task is as relevant today as it was in the 1960s and 1970s, no matter how changed the geopolitical circumstances, nor how incomplete the earlier formulations may have been. Before assessing the pitfalls as well as the persisting relevance of dependency thinking, it is important to consider, albeit briefly, the specificity of its geopolitical setting.

Turbulent Times: Contextualizing Dependency Thinking

As was mentioned in chapter 3, the 1950s and 1960s were characterized by rapid political change and instability. Against a backdrop of the Cold War, the process of decolonization, the emergence of new nations, the spread of national liberation movements, and the diffusion of Marxist thought all combined to generate a *zeitgeist* of turmoil and insurgency. In Latin America, the Cuban Revolution was a crucial marker for dependency perspectives since it raised the possibility of the emergence of authentically independent governments that would place national sovereignty above any accommodation with the United States. At the same time, US hostility to the Cuban Revolution, the quickening pace of US military aid to the armed forces of Latin American countries, and the growth of foreign investment stimulated the development of critical ideas on domination and dependency. In other parts of the Third World, especially in Africa, the decolonization process also provided a fertile ground for the rise of oppositional thinking, and the emergence of new nation-states led to debates concerning the nature of sovereignty and the impact of new forms of Western dominance, frequently represented as 'neo-colonialism' (Amin 1973).

While it needs to be emphasized that geopolitical turbulence did not of itself determine the orientation and trajectory of oppositional thinking, it did provide a significant context in which new forms of opposition to Western power would assume increased relevance. Events such as the Vietnam War of the 1960s and early 1970s, and in the Latin American context, the Cuban Revolution of 1959 and the US invasion of the Dominican Republic in 1965 – i.e. war, revolution, invasion – were emblematic of a shifting geopolitical terrain. This shifting terrain was also associated with an increased penetration of Latin American economies by US corporations. This was illustrated by the fact that by 1965 US capital represented as much as 75 per cent of all foreign investment in Latin America, compared to a figure of 20 per cent in 1914 (Sigmund 1980: 20), and that in the period from 1950 to 1970 the value of direct US investment in Latin America increased by almost three times (Niess 1990: 206), a phenomenon catalogued by Cardoso and Faletto (1979), Galeano (1973), Quijano (1968) and others.

In this overall context, a fertile basis was created for conceptualizations that challenged the Western ethnocentrism of modernization theory and sought to go beyond the established parameters of North–South relations. In this sense Hobsbawm's (1994) interpretation of the Third World as a zone of revolution and acute political instability may be amplified to include the point that such a portrayal is also relevant to the understanding of the emergence of insurgent thought. But such thinking, particularly in the Latin American case, was also embedded in a long history of alternative consciousness. Turbulent times constituted an important but not exclusive factor for an appreciation of dependency thinking. However, how relevant is dependency today under such different geopolitical conditions?

Re-reading *Dependencia* for the Contemporary Era

At the end of the 1990s, one of the more prominent Brazilian *dependentistas* from the 1960s and 1970s, Theotônio Dos Santos, noted that since dependency theory is supposed to be dead and buried, it is rather remarkable that it is still accorded so much scholarly attention. For Dos Santos dependency has not only been an 'intellectual adventure' but also a perspective which still retains a contemporary relevance, especially as seen in the way the foreign debt crisis reinforces one of its main ideas – namely, that the central countries capture economic surpluses from peripheral or dependent nations (Dos Santos 1998: 62).

Such a view might be pointedly contrasted with the perspective provided by Fukuyama, who argues that dependency is most appropriately seen as an 'intellectual mirage' – a Marxist-inspired approach that was quite unable to explain the successes of capitalist development in Third World countries (Fukuyama 1992: 108). Interestingly, when Fukuyama moves beyond the economic domain he stresses the importance of the desire for recognition, noting that such a desire is the most specifically political part of the human personality. It is here that a connection could be made with dependency studies, since the desire for a recognition of the rights of colonized and dominated peoples in the periphery was crucial to their orientation. On the central issue of the politics of recognition, Fukuyama, referring to Hegel, goes on to note that because of the successful 'economization' of our thinking, the concept of a 'struggle for recognition' seems strange and unfamiliar, whereas in actual fact it can be suggested that such a struggle is present everywhere and 'underlies contemporary movements for liberal rights' (Fukuyama 1992: 145–6). But the problem with this formulation is that it equates the struggle for recognition with one kind of motivation, the struggle for 'liberal rights', and neglects other equally important struggles against foreign domination, racial and gender discrimination, as well as popular struggles for democratic rights. However, although Fukuyama's emphasis on struggles for recognition is limited by its neo-liberal framing, the politics of recognition, as signalled in the section above, must be a central theme in any reconsideration of dependency thinking.

What I would highlight here therefore is that the struggles within the countries of the Third World for international recognition, dignity and equality were a significant driving force for much of the literature that came under the broad heading of dependency. Further, it has often been the case that this particular, sometimes only implicit, but, I would suggest, defining aspect of the dependency intervention has been given relatively little attention. I shall return to this suggestion below, but first it is necessary to recall how dependency was customarily depicted in the West, before introducing some of the key criticisms that emerged from the late 1970s onward.

As is well known, the Latin American literature on dependent development and imperialism had a significant impact on Western thought from the mid-1960s on. Specifically, the work of André Gunder Frank (1967, 1969) on capitalism and underdevelopment, a contribution which summarized a good deal of the original Latin American analysis, was particularly influential within the universities of the First World. Students read about metropolis–satellite relations, the continuing extraction of surplus by international capital, and the notion that Latin America had been capitalist since its integration into the world capitalist system through colonial conquest. This association of conquest with capitalism was a position that had already been developed in considerable detail, and published in 1949, by Sergio Bagú (1992), whose work on colonial capitalism in Latin America was never translated into English.[10]

Many of the more original themes within dependency writing – for example, the more socio-cultural aspects of dependency, as depicted by Quijano and Stavenhagen, or the more political and philosophical dimensions as portrayed by Ianni and Zea, or the more multi-dimensional and theoretical-empirical work of Cardoso and Faletto – tended to receive less attention in the West than the work of Frank. Because of this concentration on Frank, representations of dependency tended to become somewhat mechanistic and more limited than would have been the case if the full range of dependency writing had been taken into account and critically scrutinized.

By the end of the 1970s, a series of interrogations of dependency positions had been published in the West, and in the discussions that evolved it tended to be the case that the more directly economic and social aspects of dependency thinking received far more critical evaluation. Warren (1980), in a well-recognized work, condemned the *dependentistas* and their allies for not understanding the positive dynamic of capitalism. Radical Third World nationalists were also criticized for not recognizing the historically progressive nature of capitalist expansion, and for not welcoming investments by multinational corporations. For Warren these investments not only brought employment to the periphery, but more crucially laid the basis for the development of the Third World's proletarian class which would be essential for the success of future socialist revolutions.[11]

Nigel Harris (1986), in a similar intervention, argued that dependency perspectives had not been able to account for the continued economic growth of the newly industrializing countries, especially those in East Asia. He agreed with Warren that the dependency writers had presented a static view of development which had more to do with sustaining a certain brand of nationalism, rather than dispassionately analysing the realities of growth under capitalism. At approximately the same time, Lipietz (1987) in a more nuanced analysis of the crises of global fordism, also targeted what he referred to as the dogma of the 'development of underdevelopment' idea, and the prevalence as he saw it of a 'pessimistic functionalism'. Equally, however, he pointed to:

a) the unevenness of Third World industrialization; to the fact, for example, that Hong Kong, Singapore, South Korea and Taiwan accounted for 60 per cent of the South's exports of manufactures, a figure rising to 70 per cent if Brazil and India were also included;

b) the growing impact of the debt problem, whereby debt repayments involved a very sizeable transfer of resources from South to North;[12]

c) the growth in poverty and inequality within the Third World as a whole; and

d) the need, *contra* Warren, to recognize and support the struggles of workers, peasants and women for better living conditions under regimes of capitalist production.

Further to the pertinent criticism that many of the dependency writers underestimated the dynamics of capitalist accumulation, other authors argued that dependency shared many of the assumptions of modernization theory. For example, in an analysis of post-socialist trans-formations, the radical German economist Altvater (1998: 593) con-tended that while the theoretical and political orientation of modernization and dependency theory were not reconcilable, neverthe-less the normative objective of an efficient market economy, of a rich civil society and a functioning democratic political system are 'unanimously shared by both approaches'. The idea that modernization and dependency theory share a certain commonality also surfaces in Rist's (1997) work.

Rist contends that although the *dependentistas* elucidated the mech-anisms of underdevelopment and its association with class interests and the international system, they did not challenge the fundamental assump-tions of that system. Since the radicals also accepted the necessity of growth, was not their final aim also modernization, industrialization and the capture of foreign markets, and are not 'anti-capitalist' strategies therefore not also compelled to 'promote bourgeois values such as eco-nomic rationality, efficiency, utility and hard work' (Rist 1997: 121)? It is true as Rist argues that the ecological and in some cases the cultural dimensions of development were left out of account in dependency thinking, but there is little evidence to suggest that the *dependentistas* as a whole embraced the same 'bourgeois values' as the modernization theorists, also arguing for the capture of foreign markets for capitalist development.

Rist's standpoint is similarly expressed in Keith's (1997) study of international development, where it is observed that not only was dependency theory essentially economic in nature, but that the *depen-dentistas* had little criticism of the 'environmentally, psychically, and humanistically corrosive features of capitalism'; or, in other words, it would seem that they gave substantial support to the 'development-by-growth' model emanating from modernization theory (Keith 1997: 35). While it is correct to highlight the fact that dependency thinkers showed a lack of concern for environmental issues, as had the modern-ization theorists, and a lack of analysis of the psychological scars of poverty under capitalist development,[13] to suggest that their arguments replicated modernization theorists' views on growth and development, Rostowian-style, is to construct a caricature. Clearly, the main thrust of the more critical dependency positions was to challenge the international relations of power within which capitalist development took place.

The *dependentistas* argued against the subordinating impact of foreign corporations on Third World economies, interrogated the role of the internally dominant social classes within peripheral societies, opposed the dominated position of peripheral states within the international system, and in some cases supported the radical struggles of workers and peasants against existing relations of exploitation, with attendant calls for revolutionary transformation. Also, it was not that the *dependentistas* were against growth or industrialization, but rather that they were interested in the social conditions and relations under which growth took place, as well as the politics of surplus utilization. These differences are not just the residual variables of otherwise comparable perspectives. They constitute the very substance of theoretical and political divergence and conflict. Moreover, as was argued previously, dependency thinking, or a key distinctive strand within it, expressed an opposition to externally imposed imperial strategies and the governing representations that accompanied such strategies. A developing politics of identity and difference expressed a series of positions on international relations that were in sharp contradistinction to the politics of Western modernization.

Rist (1997: 118–20) raises three other criticisms of dependency which also find considerable resonance in the development literature, and which are important to consider.

First, we have the observation that it is quite difficult to identify the degree of external dependence of a country, and surely, he suggests, it is the case that all countries have been dependent at some stage in their developmental trajectories. The problem with this argument is that it conflates different kinds of dependence, and renders opaque the specificity of dependency relations as they have evolved in a First World/Third World context. Of course, as Lipietz, for example, convincingly showed, the Third World is characterized by fragmentation, heterogeneity and enormous socio-economic, ideological and cultural diversity. Notwithstanding these realities, one can still sustain the argument that imperial encounters and the geopolitical repercussions of a history of Western invasiveness have created a complex of relations between First World and Third World that are of a specific and highly significant nature.

In the international arena the societies of the periphery have far less capacity to effect changes within the societies of, for example, Western Europe and North America than vice versa. The societies of the periphery are far less able to ignore the capacity of the West – acting multilaterally through organizations such as the IMF, the World Bank, the G-7 (now G-8), the WTO – or unilaterally through Washington Administration initiatives – to effect the policies and decision-making processes within their own territorial domains. Conversely, the societies of the West have a greater degree of freedom to exclude, marginalize or 'contain' Third

World countries. In other words, what I would argue is that, overall, there is a crucial, historically rooted asymmetry of power relations as between First World and Third World in which one group of countries, the Third World, tends to remain in a dependent position. This asymmetry does not by any means preclude the development of South–South relations, but it does continue to represent a significant feature of today's international relations, a theme that will be discussed again in chapters 7 and 8.

Second, Rist reminds the reader that the *dependentistas* were strongly inclined to oversimplify international economic relations, allocating a disproportionate importance to the idea that the development of the centre was based on the underdevelopment of the periphery. It is clearly justifiable, especially with reference to the well-diffused work of Frank, to criticize what often seemed to be an overconcentration on the negative influences of foreign capital and external power in general, without at the same time considering the activities and effects within the Third World of the dominant social classes. It was also the case, and remains so today, that flows of trade and investment have been predominantly located in the economies of Western Europe, North America and Japan, and the notion that the underdevelopment of the periphery was necessary for the development of the centre was known to be a counterfactual proposition. However, what does remain clear is that a series of asymmetrical relations were created in the formation of the international capitalist system and that these relations of inequality have been maintained and accentuated into the present period, as was indicated in chapter 1.

Third, Rist indicates that the *dependentistas* did not specify any concrete solutions to the problems of dependent capitalist development, and that they were further unable or reluctant to identify the social agents who might act as bearers of an alternative model of societal transformation. Pointing to the Cuban Revolution was not a sufficiently convincing position when looking at societies like those of Brazil and Mexico, and Che Guevara's failed project in Bolivia underlined the cardinal importance of socio-political, cultural and historical specificities. The fading potential of revolutionary transformation in Latin American societies, and the return of military rule (for example, Brazil in 1964, Chile in 1973, Argentina in 1976 and Bolivia in 1980),[14] reflected the failure of radical socialist ideas to bring about a move away from dependent capitalism.

The fact that the bearers of an anti-capitalist project did not seem to be obviously identifiable was, as Rist suggests, an important deficiency

of the dependency vision. Furthermore, in many cases, a rather traditional Marxist approach to class and political action led to a restriction
of the analytical field, so that the complex questions of social subjectivity
and mobilization tended to be reduced to the presupposed historical
primacy of class struggle.[15] As with modernization theory, there was
little if any discussion of gender (Scott 1995), and, as also with modernization, the concept of 'development' remained unproblematized, in
contrast to contemporary 'post-development' currents. Despite these
and related criticisms concerning the tendency to overprivilege economic
factors, and the recurrence of ambiguities relating to the role and function of the nation-state in peripheral capitalist societies, it can still
be argued that, in important ways, dependency continues to retain
contemporary relevance. Why?

1. First, as Castells and Laserna (1989) indicated, in their overview of
changes in technology and socio-economic restructuring in Latin America during the 1980s, the worsening of the social and economic situation
in this part of the periphery was most effectively explained by the
combination of new and old forms of dependence. The new forms have
been connected to revolutionary changes in the application of information technology in the system of production, and the old forms have been
expressed through financial dependence and the imposition of policies of
austerity by foreign capital. This 'old' form of dependence has been a
characteristic of the Third World as a whole – so that, for example, and
as mentioned in chapter 4, while the foreign debt of Third World countries stood at US$67.7 billion in 1970, by 1985 it had risen to over $700
billion, representing as much as 33 per cent of the developing countries'
GNP. With respect to what Castells and Laserna refer to as the 'new'
dependence, there were, as they point out, important differences across
the periphery, so that, as is well known, in parts of East Asia, high-tech
industrialization did not lag as far behind the metropolitan countries as
was the case with most countries in Africa and Latin America. In Latin
America, for example, technological dependence related to metropolitan
control over patents and 'know-how' as well as to the fact that peripheral
states tend to invest relatively little in research and development – in
1980, for instance, while Latin America accounted for 8 per cent of the
world's population, and about 5 per cent of the world's GNP, it only had
2.5 per cent of the world's scientists and accounted for only 1.8 per cent
of world expenditure on research and development (Castells & Laserna
1989: 6). Partly as a consequence of this paucity of investment, as well as

differences in salaries and career opportunities, Latin America witnessed a significant brain drain, particularly in the field of science and technology, with the United States being a major recipient – and such flows naturally reinforced the phenomenon of technological dependence. (For a related analysis of the telecommunications sector, see Hills 1994.)

2. Second, and remaining within the more directly socio-economic realm, the importance assigned within dependency writing to marginality, inequality and poverty has lost none of its relevance (Hinkelammert 1999). For Latin America, the available statistical evidence, notwithstanding unevenness across countries and variations in time, points to increases in poverty, income inequality and informal-sector activities, especially for the 1980s (Berry 1997, IDB 1998, Lustig 1994, Tokman 1994). For large parts of Africa and India, Chossudovsky (1998) indicates the extent to which the spread of 'famine zones' has become a striking feature of the socio-economic landscape during an era of the 'globalization of poverty'. In the language of the radical *dependentistas*, it could be stated that the continuing spread of capitalism has brought with it increased poverty, inequality and marginalization, and some of the more trenchant critiques of the 'neo-liberal globalization' of the 1980s and 1990s employ a conceptual arsenal that is not so dissimilar from these earlier expressions of radical analysis (see chapter 8).

3. Third, the contemporary process of neo-liberal globalization has been accompanied by a denationalization of peripheral economies that reinforces many of the ideas developed by the dependency writers. Rocha (2002), in examining the Brazilian case during the Cardoso presidency, points to a recent study which shows that between 1995 and 1999, there were 1,233 mergers and acquisitions in which multinational corporations acquired control or achieved participation in Brazilian industries, and in the same period many a traditionally powerful Brazilian trust disappeared. This went together with a fall in the local production of capital goods, an overall growth in foreign debt and a highly unequal distribution of income, with ECLAC noting that, in 1999, Brazil was the only country in Latin America in which more than half of the population had less than 50 per cent of the mean income (qtd in Rocha 2002: 30–1).[16]

4. Fourth, connecting the socio-economic with the political and military, and foregrounding issues of national sovereignty and autonomy, Bienefeld (1994) and Blaney (1996), in different but complementary ways, highlight a series of contemporary elements of the global scene which

re-affirm key strands of dependency thinking. For Bienefeld, the new world order notion of the early 1990s ushered in a new age of imperialism where a hegemonic power, the United States, has had a greater ability and willingness, in a post-Cold War era, to override national sovereignty for reasons of its own making. Pre-9/11, the invasion of Panama in 1989, air strikes against so-called 'rogue states' such as Libya, Iraq or Afghanistan, and the continuing illegal embargo of Cuba, are just some of the actions that reveal a hegemonic power that is free to employ coercion in accordance with its own self-defined national interest (see chapter 7). In this context, as Blaney (1996) usefully reminds us, one of the most valuable arguments of the dependency writers, even if not always explicitly elaborated, concerned the *ethics* of international encounters.

Questions of sovereignty and autonomy are as relevant today as they were during the 1960s and 1970s, probably even more so, and one of the most significant facets of dependency writing was that these sorts of issues were firmly placed on to the geopolitical agenda. As a feature of the struggle for recognition, also mentioned above, the re-assertion of the sovereign rights of Third World states formed part of a project of counter-representation *vis-à-vis* the conventional tenets of modernization theory. This struggle for recognition is still relevant for a contemporary world in which international conflicts over the status of national autonomy and the power of the nation-state remain of fundamental geopolitical significance.

5. Finally, in the current era, issues of identity and difference have become increasingly significant, and one of the central aims of the *dependentistas* was to express an autonomous intellectual identity which challenged Western representations of development and change in the periphery (Fals Borda 2002). Although much of the writing took on a sometimes restricted socio-economic orientation, the geopolitical and philosophical underpinnings of the new forms of theorization represented a significant challenge to the hegemonic perspectives of Western social science and public policy in the 1960s and beyond. Dependency played its part in the construction of an alternative geopolitical memory, contesting the governing representations and ruling memory of Western metropolitan theory. The perspectives that were developed were sometimes employed by governing regimes in the South, and national elites sometimes embraced dependency ideas in attempting to consolidate their power within societies of the periphery. Dependency, like democracy or decentralization, was not a concept bestowed with a pre-given political

consensuality. It lent itself to different interpretations and emplacements; for example, from a belief in the adequacy of international economic reforms and policies of income redistribution to calls for socialist revolution. Its lasting relevance has much to do with the struggles against Euro-Americanist interpretations of global politics, and also with the vital currents of political opposition to the unilateral deployment of Western, often specifically US power, in a divided and turbulent world.

How then might these ideas relate to a post-colonial perspective, and how might dependency and post-colonial analysis complement each other in a mutually beneficial dialogue?

Dependency and Post-coloniality: Towards a Symbiotic Encounter

It is certainly the case that dependency writers tended to underestimate the dynamics of capitalist development, that they often did not sufficiently take into account the heterogeneities of peripheral societies, that the power of external factors was perhaps over-emphasized, that the complexities of social subjectivity and agency were not given full cognizance, that the possibilities of autonomous development were overplayed, and that societal transformation through revolutionary upheavals was a much more problematic pathway than was sometimes implied. It is also the case, as the Chilean sociologist Osorio (1996) suggests, that if dependency thinking is to be treated with the intellectual respect it deserves, many of its central themes and starting points need to be critically reread, so that its helpful interpretive guidelines can be refurbished and its limitations overcome.

One of the main problems facing such an enterprise, certainly in the West, has been the narrow and sometimes skewed and hostile way dependency ideas have been represented and discussed. This point was originally made by Cardoso (1977), and extended, a little over a decade later, by Kay (1989). Their criticisms of the often simplistic portrayals of dependency perspectives remain valid.[17] This is partly a factor of language and the limited dissemination of the Latin American literature, but by the mid-1990s many translations and detailed surveys were available. What I want to argue is that dependency perspectives can be usefully revisited and seen as being a part of a broad reservoir of critical thinking. I have indicated above that not only does dependency fall into a longer lineage of critical thought than is often assumed, but that it possesses a greater degree of interpretive multi-dimensionality than is frequently realized. The relevance of its multiple nature can be illustrated in two examples.

First, the emphasis given to the paramount questions of sovereignty and national independence, and to the politics of identity and recognition, especially worked out by, for example, Ianni, Dussel and Zea, still retain a contemporary resonance, as will be seen in chapters 7 and 8. Second, the highlighting of issues of global inequality and centre–periphery differentiations, either exemplified in relation to the accentuation of debt dependence or in terms of the asymmetry in decision-making power within international organizations, continue to remain themes of contemporary significance.

But, as observed above, dependency as an important critical interruption to Western thinking on modernization and development has also been limited by a number of shortcomings, which, from a post-colonial perspective, can be summarized into two main points.

1) In their critical challenging of Western modernization theory, dependency writers tended to focus on two major issues: first the unsatisfactory way the Third World was represented, with its ethnocentric essentialization of features such as 'tradition', 'backwardness', 'lack' and 'stagnation'; and second, the uncritical portrayal of First World/Third World relations whereby the modernization theorists assumed that these relations were beneficial to both parties, and in which the history of phenomena such as slavery, colonialism and imperialism were left out of account or, in the case of colonialism, represented positively. What was absent from the dependency critique, however, was any examination of the way the First World itself was depicted by the modernization theorists. The dependency writers did not extend their critique back to modernization theory's representation of the First World itself, with its predominantly unquestioning view of occidental progress, civilization and modernity. The dark side of Western history inside the West itself was not subjected to scrutiny, and in this sense the *dependentistas* failed to examine that other 'inside' – modernization's own heartland – the social foundation on which its projections for other ostensibly less-advanced regions of the world were built. In contrast a post-colonial perspective would tend to take into account the imbrications and intersections of inside and outside, of internal and external, as for example was seen in chapter 2 with the interconnections between the politics of internal US territorial expansion and the framing of foreign

policy, or in chapter 3 with the imbrications of the domestic and international dimensions of Cold War politics.

2) Second, a post-colonial perspective would bring to dependency a richer analysis of subjectivity, of identity and difference and the politics of recognition. Instead of being limited by the centrality of class analysis and capital/wage-labour relations, the analysis could be broadened to embrace the complexities of a post-structuralist approach to agency, discourse and difference. This would not necessarily entail the abandonment of the importance of materiality, of political economy, and the sharpening of global inequalities; rather these phenomena would be reproblematized, as will be suggested in the following chapter. Similarly, while sovereignty and autonomy would still figure as of central importance in a post-colonial approach, they would be situated in a context that seeks to break away from a deterministic take on the fixed, pre-given class nature of states. Revolutionary movements would not have to be seen in class terms only, nor as carriers of the violent overthrow of existing states, but perhaps more as attempts to subvert existing orders and representations, as alternative ways of doing politics, through, for instance, stressing the need for a deepening of democracy and a recognition of indigenous rights. In this context, it is worthwhile recalling that the Zapatistas bring to the political arena a different philosophy of 'revolution' and 'insurgency' while also retaining and re-inscribing ideas from the dependency 'tradition' (see chapter 8).

Turning the relation around, it can be argued that a post-colonial analytics would benefit from dependency's foregrounding of international inequalities, of disparities in the possession of decision-making powers, of the reproduction of structures of domination and of the vibrant history of critical thinking rooted in the periphery – Dos Santos's 'intellectual adventure'. There is space for a symbiotic encounter, an encounter which would be part of a project of recovery and renovation.

To explain this idea in further detail we need to move on to the next chapter, which discusses newer forms of critical thinking, from the post-modern to the post-colonial, forms which have emerged and developed since the zenith of dependency.

6

Exploring Other Zones of Difference: From the Post-modern to the Post-colonial

'Celebrating difference as exotic festival ... is not the same as giving the subject of this difference the right to negotiate its own conditions of discursive control, to practice its difference in the interventionist sense of rebellion and disturbance ... '

– Nelly Richard (1995: 221)

Opening Ideas on the Post-modern

The post-modern has frequently been associated with a 'state of mind' or analytical sensibility that carries with it an elastic, enigmatic sense of presence. In this spirit, the post-modern has remained a site of continuing debate, not least in relation to its genealogy and potential to go beyond Western frames of interpretation.

In relation to the origins of the term, it can be suggested that a range of putatively post-modern themes – the dissolution of legitimized narratives, eclecticism, readers' participation, pluralism – had already emerged in the late nineteenth-century literature of Central America (Zavala 1988). More recently, Perry Anderson (1998) has reiterated this point, noting that the term 'post-modernism' was born in a 'distant periphery', coming into European usage in the 1930s through the work of the Spanish writer Federico de Onís. Thus, as significant critical debates on development and change were initiated in the Latin American periphery through *dependencia* perspectives, equally with the post-modern, a critical history of the concept can lead us back to the Latin

American periphery. But does this mean that the signifier 'post-modern' takes us beyond the limits of Western ethnocentrism?

For Robert Young (1990: 19), post-modernism could best be defined as European culture's awareness that it was no longer the unquestioned and dominant centre of the world, a view which finds a parallel in Murphy's (1991: 124) comment that post-modernism is also 'post-colonial in its mentality'. But how far can we accept the idea that the 'post' in the post-modern is the 'post' in the post-colonial? Does the politics of the post-modern signify the end of Western ethnocentrism? With the advent of post-modern thinking, with its emphasis on plurality, difference, heterogeneity and unpredictability, is Euro-Americanism in eclipse?

My intention in this chapter is as follows. First, through a specific reading of some of the texts of key Western writers on the post-modern, I shall argue that Western ethnocentrism or Euro-Americanism as discussed in chapter 1 has not been effectively transcended. Next, I shall indicate, on the basis of a range of Latin American writings, how the post-modern has been variously interpreted in the Latin South and how these different interpretations can broaden our understanding of North–South relations. Such an understanding can be connected to the theme of the politics of difference and recognition, which I examined above in relation to dependency thinking. I shall then discuss certain aspects of the politics of the post-modern, before moving to a consideration of the post-colonial. In the treatment of the post-colonial I shall consider the contested relevance of Marxist thought and finally suggest how the post-modern differs from the post-colonial. This will then provide me with a link to the next chapter, which deals with post-colonial questions for global times.

Through this discussion, which follows on from the previous chapter, I intend to assess the emergence, characteristics and relevance of those more contemporary currents of critical thought that have influenced the way North–South relations are presently envisaged.

The Persistence of Euro-Americanism in Post-modern Thinking

As one way of illustrating the continuity of Western ethnocentrism in post-modern thought, I want to briefly discuss the texts of a number of influential writers, such as Rorty, Vattimo, Lyotard, Baudrillard and Jameson, who have been closely associated with the post-modern turn.

I shall concentrate on those ideas and interpretations that go to the heart of the link between Euro-Americanism and the post-modern intervention in social and political theory.[1]

One of the recurrent themes in the analysis of West/non-West relations concerns the ways in which the West has been constructed as a self-contained entity, and as a universalist meeting point for other particular societies and cultures. At the same time, this projection of a self-contained vision of the West has frequently gone together with a blurring of the significance of imperial politics in the production of the West (Mohanty 1992). In the domain of Western philosophy this critical observation can be further elaborated in the light of a number of contributions from the North American philosopher Richard Rorty who has been associated with post-modern ideas.

In an important exchange with the anthropologist Clifford Geertz on issues of ethnocentrism, Rorty (1991a: 13) suggests that it is only ever possible to go beyond our own 'acculturation' if our culture contains splits which can be caused by 'disruptions from outside' or 'internal revolt'. Rorty goes on to assert that 'our bourgeois liberal culture' takes pride in 'constantly enlarging its sympathies' based as they are in a 'tolerance of diversity' (Rorty 1991a: 204). In addition, the reader is informed that the majority of the world's inhabitants do not believe in human equality, which Rorty sees as a 'Western eccentricity'. Subsequently, in a chapter on Jean-François Lyotard, an important reference is made to force, whereby Rorty notes that while we Western liberals have had the Gatling gun and the native has not, equally, Western liberal societies have produced the social scientists who have shown how violent and hypocritical the West has been (Rorty 1991a: 219). There are three points that can be raised in relation to Rorty's argument, points which are even more relevant in today's world than they were in the early 1990s.

1) Referring to the non-West, it can be suggested that resistance to Western power, through, for example, movements for decolonization, has been a rather crucial and *global* part of the struggle for 'human equality', contrasting with Rorty's idea that the struggle for human equality is only a 'Western eccentricity'. Non-Western struggles for human equality and dignity could be taken, in Rorty's own terminology, as one possible 'split' or 'disruption from outside' which could help us transcend the limits of our own Western acculturation, perhaps embracing the Cuban anthropologist Fernando Ortiz's (1995) term of 'transculturation', which underlines the enabling aspects of mutual respect and learning across a cultural divide. Moreover, the posited

equivalence of the West with a unique belief in human equality is divorced from the geopolitical record of the Western diffusion of structures of *in*equality.

2) Symptomatically, in relation to the analytical limits of Occidental enclosure, critical understanding of the West's colonizing force is safely relocated inside the West itself. Whatever violence and humiliation has been associated with the colonial encounter, at least (for Rorty) the West has produced the intellectuals who are able and ethically motivated to reflect on such encounters. What is erased by such an interpretation is the presence of other agents of knowledge, located in the periphery, who have also critically analysed the nature of colonial and imperial power, as for example we have seen in the previous chapter. Furthermore, since these critiques have come from outside the Western liberal community, they have often been able to disrupt the inner images and representations of that community, and as a consequence of such a disruption we are placed in a better position to not only re-evaluate Western interpretations, but also to develop more authentically global perspectives.

3) Rorty argues that a commitment to justice, including the development of the principle of religious toleration and the institutions of large market economies, is not only an attractive feature of Western culture but also the best we can aim for, with American democracy being taken as the embodiment of all the best features of the West (Rorty 1991a: 209–11). But by advocating loyalty to the institutions of Western liberal democracy, the potentially enabling splits and disruptions from outside, mentioned by Rorty, will tend to be much more easily foreclosed. In such a contextual setting the potential for a radical questioning of Western liberalism, or the illumination of the 'nocturnal face' of Western society, will be constrained (Derrida 1992).

What is striking in these orientations, which are repeated in the 1990s, is the way a model of the West is given a positive essentiality, whereas in contrast, Rorty comments that there are lots of cultures we would be better off without (Rorty 1999: 273–6). This view is associated with a rather dystopian position that conveys the notion that perhaps there are no longer any initiatives that will save the southern hemisphere (Rorty 1999: 226). Although this standpoint may be contrasted with an earlier, more positive statement that future political hope may well lie in the imagination of the Third World (Rorty 1991b: 192), overall, Rorty's perspective does not provide a sufficiently strong basis to enable us to go beyond the limitations of Western ethnocentrism. In fact, many of his propositions help to reinforce a sense of the posited supremacy of the occident.

Remaining within the sphere of philosophy, the Italian theorist Vattimo (1991) has argued that the spirit of the post-modern is more effectively captured through seeing it in terms of the dissolution of newness and of the idea of history as a universal process. How then did Vattimo relate his notion of the dissolution of universality to Westernization?

In one important passage, it is argued that the idea that the history of Western reason is somehow the exodus from myth is also a myth. For Vattimo (1992: 42), the demythologization of demythologization can be taken as the 'true moment of transition from the modern to the postmodern'. At the same time, it is suggested that in a world of disenchantment and dissolution, where there are no longer any foundations and where there are many different cultural horizons, there is more hope for the possibilities of reciprocity and equality – for a democratic 'heterotopia'. Vattimo explores this vision by giving importance to being, envisaged in terms of dialogue and interpretation, rather than stability, fixity and permanence. The experience of oscillation in the post-modern world gives us the opportunity to find a new way of perhaps at last being human (Vattimo 1992: 11).

Along another route, however, the rather conventional association of the Third World with so-called primitive societies and the representation of underdeveloped countries in terms of lack, display little evidence of a spirit of destabilizing the customary meanings attached to the West/non-West split. For Vattimo, other non-Western cultures are seen as having found ways, supposedly paradoxical, irrational and caricatural, but also 'authentic', of entering our Western universe, so that the non-Western world is an 'immense construction site of traces and residues' (Vattimo 1991: 158). The disappearance of alterity occurs as a condition of widespread contamination, so that what Vattimo sees as the increased sameness of the world takes on a weakened and contaminated form.

Even though Vattimo does indicate that change is possible through a rather vaguely scripted process of increased cultural interaction, there is little appreciation of the fact that within non-Western traditions there are ways of thinking and analytical reflection which can provide a disrupting understanding of West/non-West interactions. In addition, little if any distinction is made between so-called primitive societies and the Third World, as if the heterogeneity of the latter can be encapsulated under the rubric of the 'primitive'. Apart from a palpable lack of curiosity in non-Western thought,[2] no attempt is made to puncture the Western position of being the world's centre of reflection and philosophical development. There is no attempt, as Rabinow (1986: 241) once expressed it, to 'anthropologize the West' and show how particularistic and exotic its construction of reality has been. In contrast to Rorty there is much more critical analysis of the West in Vattimo, but both writers appear unable to go beyond the Western notion of theory being essentially an occidental property. In both cases therefore I would argue that the post-modern sensibility does not extend to any hint of a destabilization and questioning of the centrality of Western reflection. They do not break from the Hegelian tradition of philosophy, mentioned in chapter 1, where a binary divide is posited between peoples with history and peoples without, and where the Hegelian principle of thought and the universal uniquely resides inside the West.

Similarly with Lyotard, the French philosopher most closely associated with the post-modern turn, the Third World is not seen as a place for theoretical reflexivity. In Lyotard's (1986) treatment of the 'post-modern condition' the Third World is only present in the analytical margins, or sometimes as a shadowy space for the specification of Western identity. For example, on science, Lyotard remarks that since the time of Plato, the languages of science and of ethics and politics both stem from the same perspective or choice – the 'choice called the Occident' (Lyotard 1986: 8). This suggestion connects to a contrast Lyotard makes between 'traditional' and 'developed' knowledges, the former being illustrated through a reference to the narrative of the Cashinahua Indians, with their initiation ceremonies, rituals and monotonous chants, a kind of knowledge akin to 'nursery rhymes', with the linguistic innocence characteristic of all primitive peoples (Lyotard 1986: 20–2). Although there is a fleeting reminder of the cultural imperialism of Western civilization (p. 27), the tendency to essentialize the 'traditional' or the 'developing', or in another text, to posit a notion of 'savage narratives' (Lyotard 1988: 156), sits uneasily with Lyotard's declared 'war on totality' (Lyotard 1986: 82).

Moving from the domain of philosophy to social theory, Jean Baudrillard's work on the post-modern has been characterized by a vacillating perspective on the place of the South in social change. In the 1970s, Baudrillard was one of the few social theorists who questioned Western ethnocentrism (Baudrillard 1975: 88–9). However, in a more recent text, we are presented with a series of generalizations on the societies of the South which are quite removed from the earlier critical comment on Western ethnocentrism. On Africa, there would appear to be no hope, as exemplified by the fact that its politicians would seem to embody the profound contempt people have for their own lives (Baudrillard 1990: 15). On South America, it is asserted in a later text that south of the Rio Grande, beyond the US frontier, a curse begins that is represented by the 'absolute despair of conquest which has passed into the veins of the entire people' (Baudrillard 1996: 73–4).

Negative essentializations of this type run alongside other analytical sketches which are distinctly critical of the West. For example, looking at the difference between Western and non-Western culture, it is suggested that other cultures are still exceptionally hospitable, still retaining the capacity to incorporate what comes to them from without, including from the West (Baudrillard 1993a: 142). In addition, in his text on symbolic exchange and death, Baudrillard (1993b: 125–6) makes a useful link between conceptualizations of universality and the emergence of racism, arguing that the progress of humanity and culture is a chain of discrimination with which non-Western others can be stigmatized with inhumanity and nullity. For Baudrillard, racism does not belong to the pre-modern or pre-Enlightenment time, but rather is intrinsic to the project of Western progress and scientific expansion. These comments can be linked back to his critique of Western ethnocentrism from the 1970s, but they do severely jar with his other ethnocentric remarks on Third World peoples.

There is clearly a contradictory dynamic at play here in Baudrillard's shifting portrayal of the nature of West/non-West encounters and the realities of Third World culture. Acutely paradoxical and almost at times seeming like a parody of extreme ethnocentric prejudice, Baudrillard can be read in more than one way. There is no singularity of position, and perhaps the ironies and contradictions can be seen as evidence of a play with the reader (Gane 1990). These are important issues, however, and a lack of clarity combined with a ventilation of ethnocentric prejudice do not provide a particularly fruitful route for further analysis.

In contrast, Fredric Jameson, as a key figure of the post-modern turn in literary theory, has stressed the crucial interlocking of First World and Third World realities, in which social struggles in peripheral countries like Nicaragua and South Africa have been a central element in the formation of world politics. Particularly relevant to the theme of geopolitics and post-modern thinking, Jameson (1992), in a study of cinema, aesthetics and power, makes an interesting and useful distinction between the First World and Third World.

For Jameson, imperial culture engenders a geopolitical amnesia. Inside the citadels of Empire an essentialized vision of colonized lands erases the fact of conquest and represses the history of domination. Conversely, because of being subaltern, Third World peoples are induced to acquire knowledge about the First World. They are not permitted the luxury of being able to forget the everyday implications of their geopolitical condition. Jameson draws our attention to an important feature of geopolitical power relations, a feature that can be over-generalized, but one that still holds a revealing relevance. As I argued in the chapters on modernization and neo-liberalism, this geopolitical power does not only depend on economic and military capacity, on the impinging policies of international financial organizations, or on the force of military intervention; rather it is a power that has effects since it is expressed as a mode of discursive enframing. This discursive power entails putting into place a regime of truth that subaltern nations are encouraged, persuaded and induced to adopt and put into practice. A central element of geopolitical dependency is then defined as a constriction of interpretive movement. Geopolitical power relations make it difficult to operate outside a frame of truth that is mobile and potentially enveloping.

What is evident from the writings of these Western theorists of the post-modern is that there is considerable ambivalence and enigmatic reflection. There are passages containing critical thoughts on Western universalism, but also there are limiting stereotypes of Third World culture. Overall, there appears to be a reluctance or indifference towards the creation of different kinds of conceptual framings that might help us move toward a more globally sustained post-modern sensitivity. How then has the post-modern encounter been theorized in Latin America, that part of the South where the post-modern has a deeply rooted presence?

Inside the Latin South: Contrasting Dimensions of Difference

In the contemporary debates on post-modernism in Latin America, contributions have emerged from a variety of sources, from literary theory to philosophy and from cultural studies to social and political analysis. In these debates, one often encounters an ambivalent perspective on the politics of the post-modern, which is particularly evident in the work of Nelly Richard, a Chilean literary theorist.

In an early article, Richard (1987/8) argued that, although the post-modern critique of the universalizing project of capitalist modernity had been politically enabling, conversely, post-modern writing had also tended to dissolve centre–periphery distinctions. In this way, the realities of imperialist domination became re-absorbed and anaesthetized within an apparently equivalent set of other images and meanings. This ambiguity was linked for Richard (1991) to the multi-faceted nature of the post-modern, with its connection to parody in aesthetics, decon-struction in critical theory, scepticism in politics, relativism in ethics and syncretism in culture. In this context, Richard (1992) stressed the importance of the subversive role of the intellectual, as transgressor of the ordered deployment of normalizing knowledge. The intellectual must be able to open up departures from the official projects of ordering knowledge, and here the post-modern can be potentially enabling if it encourages the destabilization of a subordinating narrative of develop-ment and modernization.

On the question of Euro-Americanism, Richard (1993) suggests that theory which is constructed in the West is predominantly inscribed with a universality which is contrasted with descriptive, empirical narratives produced in the South. Alternatively, if there is a recognition of the existence of theory in the South, this kind of theory is viewed as being local and fragmented. Thus, for Richard, any periphery that is dependent on the circuits of international organization and distribution of metro-politan knowledge is faced with the theoretical challenge of coming to terms with the problem of cultural transference. How, for instance, can Latin American intellectuals take advantage of theoretical categories put into circulation by the metropolitan networks of discourse without adhering to its hierarchy of cultural power? In a preliminary answer she suggests that one way forward would be to invoke post-modernity's themes of discontinuity and fragmentation in so far as they promise to liberate us from subjection to hierarchical totalities. It ought to be possible, she goes on, to open oneself to a dialogue with the centre that could transcend the geopolitical borders created by metropolitan control. This could be done by developing our own connections with the counter-hegemonic voices of the metropolis, with those who are democratically interested in the differences of alterity.

There is in Richard's work and in that of other Latin American writers a dual attitude towards the post-modern. Post-modernism, in their read-ing, escapes any singularity of definition. On the one hand, it can be regarded as enabling in its destabilization of the meta-narratives of

Western progress and modernization, but on the other, in its tendency to dissolve centre–periphery divisions and in its silence on global inequalities in power and income, it must be regarded as collusive with the contemporary practices of neo-liberalism.

The potential link or collusion between a certain kind of post-modern thinking and neo-liberalism has been highlighted by Reigadas (1988: 142), who asks why Latin Americans should renounce collective projects, celebrate the end of utopias and ideologies, and declare that liberation is an old myth? Why should it be assumed, she goes on, that nations are obsolete, when Latin America was always prevented from constructing its own versions, and why should the historical ideals of solidarity and justice be renounced in exchange for a post-modern individualistic culture in which anything goes? In Reigadas's vision the post-modern is intimately linked to the neo-liberal, with this double occidental move signifying a growing 'cultural colonization'.

Hence, while both Richard and Reigadas concur on the need to interrogate neo-liberalism, Richard adopts a more nuanced view of the post-modern, preferring not to simply condemn it as another modality of cultural and political colonization.

The same theme surfaces in Hopenhayn's (1988) critical observations on culture and development. How, he asks, can we incorporate the post-modern debate in order to reactivate the cultural basis of development, without leading us into a post-modernism that is functional to neo-liberalism's project of political and cultural hegemony?[3] Hopenhayn provides elements of a possible answer to this and related questions by highlighting the importance of the revalorization of democracy and a change in the perception of social scientists faced by a multiple array of social actors. An emphasis given to new social movements goes together with the need to accept the search for new forms of doing politics, in which the recognition of cultural diversity figures prominently. Hopenhayn expresses the belief that it is through insights from the post-modern debate that the cultural base on which modernization in Latin America has been constructed may be better understood. More specifically, he argues that through a post-modern problematization of the 'cultural cement of modernization' it may be easier to break through the neo-liberal frame on development thinking.[4]

At the same time, the post-modern needs to be connected to the onset of political disenchantment, with the eclipse of the image of revolutionary change. For Hopenhayn (2001: 15) this political eclipse or mortality has been produced by a combination of military coups, the success of neo-liberal ideology, real socialism's collapse, the rationalization of the market and the triumph of pragmatism in the political arena. Together these phenomena undermined the idea of revolution as an attainable horizon. And yet Hopenhayn still holds out for a new vision of utopia, a kind of post-modern utopia. This vision which borrows from the enabling aspect of the post-modern, connects with the politics of *mestizaje*, a specific kind of Latin American hybridization, whereby an attempt could be made to 'negate the negation of the other' and to open the 'repressed abundance of intercultural riches inscribed in our history' (Hopenhayn 2001: 153).

The theme of *mestizaje* and cultural hybridization connects to related interpretations of post-modernity in multi-ethnic societies such as Bolivia, Mexico and Peru (see e.g. Calderón 1987, Gómez-Peña 1992 and Quijano 1988). In the case of Peru, Quijano stresses the importance of rescuing notions of reciprocity and collective solidarity, embedded in the cultural history of the Andes. These cultural practices can be seen as forming a basis for a more enabling (post-)modernity, in which ideas of the nation and of community can be rethought in the face of the neo-liberal challenge.

In the Mexican example, García-Canclini (1991: 24 and 1995) has inferred that in as far as both hegemonic and popular cultures are hybrid cultures, it is undeniable that in this sense Latin America is living in a post-modern epoch, in a time of *bricolage* where diverse epochs and previously separated cultures intersect with each other. Also in a Mexican context, Roger Bartra (1991) underscores the advisability of seeing contemporary national culture as an amalgam of popular culture and transnational mass culture; the advent of satellite television, commercial music, imported comic books and novels has become an integral part of Mexican political culture just as the 'Mexican soul' – melancholy, fatalism, negligence, violence, resentment and evasiveness – has survived the avalanche of foreign influences to retain a stable place in the nation's political culture. Moreover, with the pivotal existence of a border territory between two cultures, a key frontier between North and South, there is an overlapping of 'Mexicanization' with 'Americanization' on the connected but other sides of that border.

There is here a post-modern sense that the 'borderization' of the world can provide points for a more effective contestation of oppressive practices and official discourses. As with other Latin American observers of the post-modern, there is a persistent desire to engage with the present and past as a way of thinking a more creative future, no matter how problematic. Further, there is in the Latin American discussion both an inside and an outside. On the one side, there is both a continuing analytical engagement with theory disseminated from the West, *and* a critical scrutiny of the effects of occidental penetration on the political and socio-cultural structures of Latin American societies. On the other, there is a restless examination of the internal specificities of these peripheral societies, so that the 'inside' and the 'outside' tend to intertwine, separate and recombine. The West is not an unnamed Other, a haunting shadow. Its presence is unavoidably interwoven into the Latin South's own constitutive inside, but this interweaving includes a critical spirit of challenging the effects of the occidental presence, of calling for a liberation from the 'canons of metropolitan imposition' (Gómez 1988: 93).

Piscitelli (1988), for example, draws our attention to the idea that the posited exhaustion of modernity has been revealed in its failure to recognize the multiple alterities that have resisted its disfiguring and destructive hegemony. It was so customary, he continues, to think that the non-occidental was also the pre-rational that it became feasible to deny the fact that modern culture was always a culture of external and internal imperialism, with its 'other scene' of the Cosa Nostra, corruption and violence. But also, Piscitelli suggests that it may well be time to invert our interpretative codes and change the terms of comparison. Hence, instead of seeing the 'imperfection' of the Latin American political form in the limited and insufficient gaze of the North, Latin Americans may make their comparative analyses as an anticipation of what will be seen in the North in a few more years. In this way, the South, or in this specific case the Latin South, may be seen as 'ahead' of the North, as a world which offers to its supposedly more advanced Northern Other a picture of what the North may become. Notions of the 'Third Worldization' of the First World, the growth of 'informal sectors' and the increasing prominence of cultural hybridization point in the direction of this kind of representation with its destabilizing and innovative potential. As we saw in chapter 1, there are problems with the idea of a Third Worldization of the First World, but in this case Piscitelli's objective is to employ a subversive post-modern insight in order to question what is invariably taken for granted – that it must always be the First World that writes the future for the rest of the world, because those other worlds are always envisaged as being behind, as needing to follow in the wake of the West.

Having drawn out some of the most distinctive differences between the Western and Latin American debates on the post-modern, it is now necessary to focus more specifically on to the politics of these post-modern debates, with a key reference to the North–South dimension.

The Politics of the Post-modern

There are at least two modalities of political thinking in the domain of the post-modern. Initially, one can identify an influential current which tends to downplay the possibilities of going beyond the everyday realities of the existing political order. Lyotard (1997), for example, contends that the liberal capitalist system under which we live is not subject to radical upheaval but only to revision. In a similar vein, Baudrillard (1994) has asserted that there is no revolt any more, no antagonism, no longer any convictions, no longer any real opposition; the 'combative periphery' (e.g. the Third World) has also been reabsorbed, and all forms of concrete freedom are being absorbed into the only freedom which remains, the freedom of the market (Baudrillard 1989: 116; 1998: 55–7). Such a melancholic vision finds a parallel in Lipovetsky's (1994) idea that for the first time we are living in a society that devalues the self-denial associated with the pursuit of a higher societal ideal, and instead systematically stimulates immediate desires, the passion of the ego and materialist forms of happiness. Contemporary society is, for Lipovetsky, witness to the 'twilight of duty' and to the prevalence of a post-moralistic logic in which narcissistic individualism, the seductiveness of consumerism, an ethos of the ephemeral, and the continuing differentiation of commodity production constitute the distinguishing features of the era – an 'era of emptiness'.

Notwithstanding differences of conceptual and thematic orientation, Baudrillard, Lyotard and Lipovetsky, as symptomatic observers of the post-modern, share a view of neo-liberal times in which there would appear to be a continuing present, an 'empire of the ephemeral' and an eclipse of radical, insurgent politics. Such a deployment of the post-modern has not passed without critical comment. Guattari (2000: 41–2), writing at the end of the 1980s, argued strongly against the 'fatalistic passivity' and 'destructive neutralization of democracy' characteristic of certain kinds of post-modern thinking, while Rancière (1995) took issue with those writers, such as Lyotard, whom he considered as undermining any optimistic reading of post-modernity, especially through minimizing the continuing relevance of political struggles and a democratic ethos. There are some important issues here.

First, it is worthwhile recalling that trends associated with the post-modern, such as the proliferation of difference (as related to, for example, gender, ethnicity, religion, nationality, locality), the decentring of the social subject, the plurality of subjectivities and the end of pre-given unitary views of emancipation, do not have to usher in a mode of thinking that abandons attempts at radical reconstruction and passively accepts the neo-liberal nostrum that there is no alternative to the present disposition of power relations. In fact, challenging the neo-liberal can encourage us to think more broadly about the nature of politics, a topic that is acutely present in the Latin American literature on the post-modern.

In the work of a number of political theorists from the Latin South who have engaged with post-modern thinking – for example Aricó (1992) and Portantiero (1992), as well as Hopenhayn (2001) and Piscitelli (1988) – there has been a constant sense of involvement with all the problems of constructing a new politics that is independent and critical of the influence of neo-liberalism. In addition, there has been a clear distancing from those earlier socialist discourses that always presupposed an unquestioned foundation, or ultimate ground, from which all political meaning acquired its historical significance. The fact that Aricó entitled his paper 'Rethink Everything' gave expression to a pervasive current of critical political thought that saw the need for constructive and questioning interventions, as well as for the invention of a new political culture with new forms of collective action. It was also appreciated that this was going to be a problematic pathway and some observers were more sombre about the possibilities of new forms of radical democracy developing in a neo-liberal era (see e.g. Lechner 1991). Nevertheless, in contrast to the melancholic reason of the Western post-modern, the Latin American intellectual climate was witness to a more engaged and constructive spirit of conceptualizing socio-political change.

Second, the tendency to preclude the possibility of political alternatives – assuming that the triumph of liberal capitalism is irreversible – is itself another kind of meta-narrative, and closes off the full range of social and political possibilities which are still thinkable and open to collective action, for example the drive to extend and deepen democracy. In this context the failure of the post-modern approach is that it has transformed the awareness of the disintegration of totalizing discourses of political change, such as certain kinds of Marxist analysis based on the centrality of class, into an assertion that dispersion and disunity is the only reality of our political world. As Laclau (2000: 301) succinctly puts it, the post-modern perspective has tended to transform the

epistemological failure of classical totalizing discourses into an ontological condition of what is actually taking place in our social world.
Cynical reason – a diffuse and melancholic cleverness that prioritizes
the negative experiences of social and political life – is a feature of
post-modern thought that draws a veil over the real possibilities for a
more emancipatory politics and a society where struggles for justice and
equality are not inevitably doomed.

Third, what has been referred to as the 'oppositional' nature of post-
modernism (Santos 1999) can be enabling in that it can help us move
from monocultural to multicultural forms of knowledge, from knowledge as regulation to knowledge as emancipation, and as intervention it
can engender a move from conformist to rebellious forms of action.
None of these moves are essentially rooted in post-modern thought;
rather, as we have seen in my review of some of the Latin American
literature, they are more appropriately seen as one possible reading of the
post-modern which can be used as part of a critique of all forms of
essentialization.

There is then what we can identify as an alternative expression of the
politics of the post-modern. This 'oppositional' form contrasts with the
melancholic post-modern, which comes close to terminating the political
altogether. Moreover, an oppositional post-modernism has been reflected
in those Latin American currents that have questioned the dissolution of
centre/periphery distinctions. The questioning of Western universalism
and the argument that North–South differences are still very much part
of our globalizing world can be seen as a key dimension of a Latin
American post-modern that has not lost its critical edge.

Analytically, the survival of a critical edge to the post-modern can lead
us into posing a number of questions which are relevant to this stage of
the argument: First, does a post-modern sensibility entail the abandonment of Marxist thought? Second, how does the 'post' in the post-
modern differ from the 'post' in the post-colonial? And third, how
might a critical post-colonial analysis connect to a reproblematized
Marxism?

Theory and the Post-colonial

In responding to these questions, I want to provide a focused clarifica-
tion that takes up some of the theoretical points broached in chapter
1. My own position is that the various modes of analysis, frames of

interpretation and conceptual lenses that we have at our disposal (for example, from Marxism through the post-modern and post-structural to the post-colonial) are most appropriately viewed as a *combined resource of critical thought*. This does not mean that one has to opt for a blurred eclecticism. A position needs to be advanced and defended, but in the spirit of critical, reflexive thinking mentioned in chapter 1, where, for example, differences and commonalities within specific theoretical traditions ought to be identified and taken into account, just as the presences and absences within particular perspectives also need to be borne in mind. Let us begin with an example from the interface between Marxist analysis and post-modern/post-structuralist thought.

It is in Jameson, one of the most well-known thinkers of the post-modern, that one of the clearest statements of the centrality of class and the mode of production can be found. For Jameson (1986: xiv–xv), writing in the mid-1980s, and again in the mid-1990s (see e.g. Jameson 1997), Marxism as a coherent philosophy stands or falls with the matter of social class. Further, the 'mode of production' is the fundamental category of Marxist social analysis which, for Jameson, would seem to remain essential for people committed to radical social change. Jameson's underwriting of a particular reading of Marxist thought that prioritizes class analysis can also be found in the work of other writers associated with post-modern theory (such as Žižek 2000). If, however, it is believed necessary to provide a critique of the essentiality of class and the economy, without abandoning Marxist thought in its entirety, how do we proceed? I would suggest that the following two points might be helpful.

First, in one early Marxist current, dating from Engels, it was posited that the economy remained determinant in the last instance.[5] In other words, it was presupposed that the logic of capitalist development governed the outcome of social and political processes, including the nature of the state. A key problem of this tendency was that together with the posited centrality of the economic structure, social subjects came to be envisaged as being absorbed within this determining structure. Such an analytical current, which represents an unalloyed form of economism, is now quite rare.

Second, in much Marxist analysis, when there has been an examination of socio-political change, the key subject has invariably been a class subject, and the class struggle has been interpreted as the defining historical struggle. There have been three interconnected difficulties here.

a) In the first place, there has been a failure to analyse the ways in which different forms of social subjectivity come into being. The varied processes through which individuals in society are formed as social subjects or agents have tended to be neglected, since the overriding concern has been with the formation of class subjects in the context of the mode of production. Instead of seeing the class category as one possible point of arrival in an examination of social subjectivities, class has been taken as a pre-given point of departure.

b) Second, it has been assumed that in the formation of social consciousness, the point of production is central and determining. The world of production and wage labour has always been the cynosure for any examination of the politics of identity and mobilization. Consequently, an understanding of the heterogeneity and complexity of social consciousness has been somewhat circumscribed. In addition, since the social subject has been interpreted as a unified subject, centred around the experiences of the workplace, it has been less possible to begin to comprehend the barriers to mobilization at this site of potential conflict.

c) A third problem has been that frequently in Marxist interpretations the proletariat, or more broadly the working class, has been conceptualized as *the* privileged revolutionary social subject, given its location in the relations of production. This has had the effect that less attention was given to the processes whereby varying forms of political subjectivity can be constituted, and through which the propensity to act collectively may emerge in specific circumstances. These processes of political subjectivity do not have to be class based – they can also be related to gender, ethnicity, nationality, race, regional identity, neighbourhood, the struggle for democracy, social justice, environmental ethics and so on. And in those cases in the past where there have been revolutionary changes in Third World societies, for example in Cuba and Nicaragua, it would be difficult to explain the mobilizations that took place as only or essentially class based. In contrast, it might be more apposite to suggest that a variety of social subjects, from a broad range of socio-economic and political backgrounds, unified at a given moment around a specific vision. This vision combined a range of attitudes, feelings, objectives and desires – around questions of the nation, of the fight against dictatorship, of the need for social justice and equality and opposition to US hegemony. United around this vision these social subjects came together as a movement with an organized leadership and took a series of actions that culminated in the moment of revolutionary rupture.

Is it possible then to identify the key agents of historical change – to make a choice between class and social movement? The problem with the

Marxist prioritization of social classes as the key agent of historical change has been that such a perspective assumes what needs to be shown. It assumes that an abstract category – class – already possesses an active agency, that it is already known and pre-constituted. Such a response bypasses a more primary question concerning the constitution of social agency.

While the kind of 'traditional' class analysis so often a part of Marxist thinking has its limitations, as noted above, it is important not to reject the Marxist perspective outright as if there are no differences within it, no concepts which are still relevant, no rereadings that are still germane. The Gramscian variant, which profiles concepts such as hegemony, as briefly discussed in chapter 1, 'war of position' and collective wills, whereby mobilization and collective action do not have to be reduced to a class category, can provide the basis for a more constructionist and less rigid approach to many central political questions. A Gramscian perspective, for example, can be used to prioritize questions of the political, and in a way that is less essentialist than other Marxist currents. Briefly, let us take three examples.

First, with the notion of 'collective will', Gramsci (1977), writing a short piece in 1917 on the Russian Revolution, and symptomatically entitled the 'Revolution Against "Capital"', observed that events had overcome ideologies – 'events have exploded the critical schema determining how the history of Russia would unfold according to the canons of historical materialism'. The Bolsheviks, for Gramsci, had a vision that saw the dominant factor in history to be, not raw economic facts, but men in relation to one another, developing a collective social will – 'men [*sic*] coming to understand economic facts, judging them and adapting them to their will until this becomes the driving force of the economy', resembling a 'current of volcanic lava' (Gramsci 1977: 34–5). Here, Gramsci is breaking away from the limits of class determinacy and emphasizing the openness, unpredictability and dynamism of collective social will, captured in the metaphor of 'volcanic lava', a magma of collective subjectivity that closely resembles the notion of the political discussed in chapter 1.

The significance of collective social will is developed in more theoretical detail in *The Modern Prince*, where emphasis is given to the importance of the development of a national-popular collective will in which issues of nationalism and the North–South division in Italian society are combined with questions of class struggle (Gramsci 1975: 135–88). Similarly, the weight of conscious will is also underlined in Gramsci's comments on liberalism, which he notes is an 'act of will conscious of its own ends and not the spontaneous automatic expression of an economic fact' (Gramsci 1975: 153). Equally, Gramsci's notion of collective social will could be applied to an understanding of the Cuban and Nicaraguan Revolutions, as mentioned above, since while the concept connects to class background it fundamentally foregrounds the importance of an articulated unity in which issues of, for instance, the national-popular, social justice, anti-imperialism and equality are linked or sutured.

Second, and intimately linked into the concept of collective will, Gramsci's notion of a 'war of position' retains a contemporary relevance. For Gramsci, political struggle is enormously more complex than military war. To illustrate this point he gives the example of India's political struggle against the British, which expressed three forms of war – a war of movement, which is the moment of armed struggle, a war of position and underground warfare (Gramsci 1971: 229). Gandhi's passive resistance is for Gramsci a clear example of a war of position which at certain moments becomes a war of movement and at others underground warfare. The war of position is the collective struggle for popular or counter-hegemony, another kind of societal leadership.

Third, and rather crucially, we have the concept of hegemony. As was noted in chapter 1, Gramsci's concept of hegemony, put very simply, combines leadership, based on consent, with the capacity for coercion. A hegemonic strategy is a strategy that rests on the ability of dominant groups in society to provide moral, cultural, intellectual and political leadership; to succeed, through persuasion and inculcation, in making their own ideas the ruling ideas of society; to generate broad consent towards their project of leadership, which is also rooted in the capacity, through the control of the state or political society, to exercise coercion if and when necessary. There is then in Gramsci a double power, a combination of leadership with force, wherein the crucial element in hegemony is that of the desire and capacity to lead through persuasion and the dissemination of a political project (Fontana 1993). Projects of hegemony and counter-hegemony are intimately linked into the importance of conscious wills and the desire to construct new political relations of power in society. The struggle over ideas and interpretations are a crucial part of these projects, as Laclau and Mouffe (2001: vii –xix) have recently indicated in their post-structuralist or anti-essentialist reading of the Gramscian perspective. And also here, it needs to be added that it was not only ideas and interpretations that were crucial for Gramsci, but also the political significance of feelings, of passion – 'one cannot make politics-history', writes Gramsci (1971: 418), without passion, 'without this sentimental connection between intellectuals and people-nation'.

Hegemony, collective wills and war of position are three concepts from Gramsci's theoretical work that point to a continuing relevance for post-Marxist thought. By this comment I do not mean to imply that Marx himself is not relevant or that new readings are not possible. Rather, given that a key emphasis of the above-cited texts falls on politics, space and power, the Gramscian point of departure concerning the theorization of the political is clearly of relevance to a contemporary critical geopolitics. It has been suggested in fact that Gramsci could be considered as an early 'post-modern Marxist'. Further, and given the fact that his theory of hegemony and subordination has been portrayed as originating 'from the periphery of modernity' (Urbinati 1998), we might ask to what extent could Gramsci be considered a 'post-colonial Marxist'? In attempting to answer such a question one finds an inescapable ambivalence in Gramsci.

For example, along one interpretive route, Gramsci (1971: 416) seems to be rather uncritical about what he calls the 'hegemony of Western culture over the whole World culture', a statement that might be connected to his negative remarks on the 'backward masses of Africa', in his discussion of intellectuals (ibid.: 21). Conversely, writing about 'war in the colonies' in 1919, he connects the colonial world with the idea that the 'class struggle of the coloured peoples' points to the 'vast irresistible drive towards autonomy and independence of a whole world, with all its spiritual riches' (Gramsci 1977: 60). Furthermore, in a *dependentista* spirit, Gramsci writes of how colonial populations are subjugated to the interests of the mother country in that they have to produce 'cheap raw materials for industry...for the benefit of European civilization'. Gramsci concludes this passage by stating that 'colonial populations become the foundation on which the whole edifice of capitalist exploitation is erected', and as a consequence these populations are deprived of the 'necessary conditions for their own autonomous development' (Gramsci 1977: 302).

Thus there are in Gramsci traces of a dependency perspective, combined in a rather dissonant fashion, perhaps, with an acceptance of an occidental centrality in world culture. Does this make Gramsci a post-colonial Marxist? Not entirely, since a post-colonial perspective would convey a greater criticality towards the Western cultural project of assimilation. At this juncture, let us stress the point that it is the difference and ambiguity within Gramsci that can be employed to help us avoid rigid judgements and hold open interpretations that recognize a place for fluidity and plurality. This can mean, in the context at hand, that there will be a variety of ways of seeing the intersections among Marxist, post-modern and post-colonial thought.

One way of providing a focus for these intersections of creative tension is to return to the previously posed questions concerning, firstly, the relation between Marxist thought and the post-colonial and, secondly, the difference between the post-modern and the post-colonial.

On the intersection between the Marxist and the post-colonial, there can be a tendency to frame Marxist thought in a way that forecloses the possibility of considering the differences within it. For example, Robert Young, in his comprehensive treatment of post-colonialism, begins by defining post-colonial theory in terms of a political analysis of the 'cultural history of colonialism', and stating that the assumption of post-colonial studies is that many of the wrongs against humanity are a product of the 'economic dominance of the north over the south' (Young 2001: 6). For Young, post-colonial analysis operates within the historical legacy of a Marxist critique on which it continues to draw, so that a post-colonial cultural critique integrates its Marxism with the politics of international rights and with a focus on how socialism can be developed in a popular rather than coercive form. In Young's reading, the problems of a Marxist analysis tend to be avoided, with the differences between Marx and Gramsci being bypassed, although there are some insightful passages on Gramsci in relation to subaltern studies (Young 2001: 353–5). In other cases, one can find assertive support for rather traditional Marxist perspectives, with, for example, the underscoring of the posited centrality of class and capital (Parry 1994: 15-16). But it is against this form of centralized theorization that post-colonial writers such as Bhabha (1994) have developed a more multi-dimensional analysis of agency and difference, in which prioritizations of the totalizing force of capital are depicted as all-too-predictably knowable.

What is at stake here is the clash of different theoretical positions within the overall domain of post-colonial analysis. This can be interpreted as a contest between conceptual and thematic persuasions that base themselves within a rather traditional Marxist problematic, founded on the central categories of capital and class, production and the world of wage-labour, and other positions which reflect a greater affinity for post-structuralist and post-Marxist thought. In this latter case, class is no longer necessarily central, and concepts of subjectivity, identity, difference, representation, agency and resistance are deployed in ways that are more open to new or different modes of enquiry (see chapter 8).

Their critics would counter that such an openness frequently fails to take into account questions of materiality, inequality and economic power. Those writers who adopt a Marxist perspective tend to argue that the post-colonial turn has frequently been associated with eclecticism, an avoidance of political economy, and in particular class politics, and more generally a failure to critique global capitalism.[6]

This kind of criticism is somewhat misplaced since, if we examine the work of the three writers most closely associated with the post-colonial as a site of critical enquiry, namely Bhabha, Said and Spivak, it becomes clear that while all three deploy hybrid perspectives, they all, albeit in different ways, borrow from the reservoir of Marxist thought. Said (1978 and 1993), for example, combines conceptual groundings from Gramsci as well as Foucault, while also being rightly critical of the tendency to turn Marxist categories into terminal abstractions. Spivak (1988, 1990 and 1999), in her work on the subaltern, representation and post-colonial literatures, combines Marxist categories with a post-structuralist sensibility, while frequently introducing such categories as 'the international division of labour', 'neo-colonialism' and 'global capitalism', in ways that convey a sense of a one-dimensional reality. Perhaps Homi Bhabha (1994 and 1996) appears to be the least Marxist of the three, but here there is much more of an attempt to reproblematize Marxism rather than to reject it as a unified system of thought (see Bhabha 1994: 27)

Overall, I would argue that in all three writers of the post-colonial turn one finds shades of theoretical meaning and a sustained hybridity of critical analysis which differs sharply from those interventions which are inclined to assign to Marxist theory a pre-given, homogeneous and privileged status.[7]

Having considered certain relevant aspects of the Marxist/post-colonial intersection, we are still left with one of our earlier questions: namely, how is the 'post' in the post-colonial different from the 'post' in the post-modern?

In contrast to a post-modern perspective, a post-colonial approach tends to be motivated by the need to focus on the centrality of the colonial/imperial interface for understanding global politics. It is not only that West/non-West encounters figure more prominently in the analysis, but that the imperiality of Western power is taken as a crucial component for understanding modernity and the global. Moreover, there is an inside and an outside to this analytical preference, with questions of ethnic/racial differences within the West being linked to the impact of colonial encounters on the nature of international relations. In other words the post-colonial critique not only refers to the colonial antecedents of modernity, but it also encourages us to rethink the limitations within the West of a consensual liberal sense of societal harmony.

A second difference can be linked to the issue of the agents of knowledge. Here the Western post-modern, as was noted earlier on in the chapter, still implicitly reflects the view that the life of the mind resides in the West and that the Third World remains too exotic for theoretical analysis. The Third World intellectual is rendered invisible by the Western take on the post-modern, whereas in contrast post-colonial writing largely affirms the significance of non-Western thought. A third contrast between the post-modern and the post-colonial relates to the question of challenging the existing disposition of power relations. Although, after Santos (1999), an 'oppositional post-modernism' can be associated with an ethical stand against injustice and oppression, it is more often the case that a questioning ethico-political position emanates from a post-colonial perspective.

Finally a further distinction between the post-modern and the post-structural can be suggested, terms which are frequently used interchangeably, underlining the importance of identity, representation and agency. The post-structural might be viewed as a theoretical pathway that seeks to go beyond a previous perspective that was centred on the posited primacy of structures acting on subjects, whereas the post-modern, certainly in its oppositional form, might be more appropriately interpreted as an orientation or analytical sensibility which seeks to go beyond the conceits of being modern with all its associated baggage of progress, unilinearity, singularity and secular reason.[8]

Concluding Comments

Having examined some of the distinguishing features of Marxist, post-modern and post-colonial approaches, and more specifically their interwoven tensions, differences and potential complementarities, it is now necessary to return to my initial outline of a post-colonial perspective as set down in chapter 1. It will be recalled that this perspective had five elements.

First, the post-colonial analytical sensibility focuses on problems of difference, agency, subjectivity and resistance, but in contrast with the post-modern and the post-structural, it does so in a way which aims at challenging and destabilizing Western discourses of, *inter alia*, progress, civilization, modernization, development and globalization. This disruption and displacement of Western ethnocentrism is effected through critically foregrounding the central importance of colonial and imperial politics. In this context the invasiveness of Western power is taken as a key phenomenon and not relegated to the analytical margins.

Second, the post-colonial can be deployed to bring into play the mutually constitutive role played by colonizer and colonized, centre and periphery, or globalizers and globalized, and this can include the points made above in relation to the stress on the overlappings of inside and outside in the domain of race and cultural difference. In terms of these complex overlappings, the emphasis will always vary as between the inside and the outside, but the salient point is that in the post-colonial there will always be an awareness of their interpretive imbrication.

Third, the post-colonial as a mode of analysis can be employed to raise questions concerning the geopolitics of knowledge. For example, who are the agents of knowledge, where are they located, for whom do they speak, what are the conceptual and thematic priorities, where are the analytical silences, who is being profiled and who is being marginalized? This can lead us into an appreciation of the significance of counter-analyses emanating from the societies of the global South, so that conceptualizations of power, space and politics are not confined to a 'Western heartland'. Throughout this text, I have sought to take account of these sorts of issues, while realizing the extensiveness of the analytical terrain that they imply. It also needs to be indicated here that it is in relation to these kinds of questions that the post-colonial is closely connected to feminist theory.[9]

Fourth, the post-colonial, as I deploy it in this text, profiles the periphery or the societies of the South as a way of making more visible the global realities of geopolitical power and representation. The advantage in looking at peripheries is that power can be clearly revealed in what it drives from the centre to the edges, so for instance US geopolitical penetrations of Third World societies, such as aggression against the Sandinista government in the 1980s or the present-day invasion of Iraq, reveal the other side of US democratic government – the imperial nature of its power. While being aware of the multiple nature of peripheries, and the pitfalls associated with radical spatialism, where, for instance, centres 'exploit' peripheries, the post-colonial can be enabling in the emphasis it gives to the geopolitically relational nature of centre–periphery encounters.

Fifth, the post-colonial carries with it an important ethico-political dimension that is rooted in the critique of colonialism and imperialism and in the revalidation of autonomy and resistance to subordination. The post-colonial, in contrast to the post-modern, as noted above, has more effectively preserved a critical spirit, and has not jettisoned its ethical positionality, seeking to go beyond the imperiality of knowledge. I shall return to this dimension in chapter 9.

These five elements have been present in differing ways and to differing degrees in the text so far, but they need to be further elaborated in relation to questions of globalization and contemporary social struggles. Moreover, although I gave a brief specification of what I understand by Empire or imperialism at the end of chapter 2, this is a component of the post-colonial critique that requires further analysis. For example, in global times, the following questions can be posed:

- Is imperialism still a relevant phenomenon?
- How does globalization connect with the debate on resurgent US power?
- How do we situate 'anti-globalization' protests on the terrain of social movements, democratization and resistance?
- What are the differences in the ways these questions are treated across the North–South divide?

These are the key questions that I shall now take up in the next two chapters.

Part IV

Geopolitics in a Globalizing World

7

Post-colonial Questions for Global Times

'While I am aware that the center/periphery distinction is suspect in this era of mass globalization, I would like to maintain it, if only because globalization has neither erased nor supplanted this distinction, which has been maintained through 500 years of Western religious, economic and cultural expansion.'

– Walter Mignolo (2000a: 183)

Re-imagining the Geopolitics of the Global

As the literature on globalization continues to expand, and most noticeably so since the mid-1990s (Taylor, Johnston & Watts 2002: 2), more attention is being given to the diversity of meanings attached to the term 'global'. The global may surface in relation to the intensified formation of flows, whereby the accelerated movements of money, images, information, migrants, drugs, new technologies and viruses are viewed as phenomena that radically transcend the territorial confines of the nation-state. In another context, the global may be linked to the strategy of a transnational corporation, or employed as an image to sell a commodity, or deployed as a key symbolic reference in a programme of political mobilization around issues of environmental degradation. The multiplicity of meaning attached to the global feeds into broad definitions of the process of globalization, which has been linked to the growth of worldwide networks of interdependence (Held & McGrew 2002: 1). This sense of intensifying global interdependence and interconnectedness, which stretches across a variety of spatial scales and which is characterized by a persistent growth in the spatial density of connections, communications and networks, is consistently present as a defining feature of many specifications of globalization. These specifications are of course very much affected by the discursive frame in which they are situated. In chapter 4, the neo-liberal frame for locating globalization was given some critical attention, and in this chapter I intend to consider certain post-colonial questions for the analysis of globalization in which the geopolitics of the global will assume a central place.

My first suggestion is that the process of globalization, fuelled as it is by 'turbo' or 'fast' capitalism, has been and continues to be configured not only by *unevenness* but also by an important series of *tensions* and *counterpoints*. Thus, while the processes of global economic integration proceed, the more trends towards social and political disintegration become accentuated. This phenomenon has been described in terms of a combined dynamic of fusion and fission (Ramonet 1997). Hence, on the one side there is a drive towards supranational economic integration, as exemplified by NAFTA (the North American Free Trade Agreement), the European Union, Mercosur (Argentina, Brazil, Paraguay and Uruguay) and the projected FTAA (Free Trade Area of the Americas), and on the other, propelled by the energies of resurgent nationalism and discourses of ethnic identification, multi-ethnic states are destabilized from below by new political fissures. Moreover, while instantaneous electronic movements of money and messages give meaning to notions of a 'borderless world', in other zones 'fortified enclaves' or 'gated communities' are erected to separate high-income spaces from the social worlds of poverty, crime and disorder (Caldeira 1996). In a similar vein, an explosion of interconnectivity is contradicted by a turning in, reflected for example by a tangible reduction in the coverage of foreign affairs by key Western media outlets (Moisy 1997). Hence 'going global' can exist side by side with a tendency, certainly visible in the US, towards the re-assertion of an inner-directed gaze, a tendency also reflected in the content of US textbooks on international relations where the world beyond the US, albeit pre-September 11, was minimally present (Dalby 2003: 148)

An ethos of turning inward can also be connected to a refusal to recognize the rights of others, be they from ethnic minorities, different religions, migrant communities or poor neighbourhoods. In the US and Europe, the desire to defend borders and erect 'fortresses' sits uneasily with support for the free movement of commodities, open economies, the abolition of economic protectionism, and deregulation. Hence, while on the one hand the opening up of space to the free flow of capital is championed, the free flow of labour is checked at the border. Within the transnational space of NAFTA, for example, the US places increasing restrictions on the inflow of Mexican labour while reasserting the centrality of open economies; in fact, for some Mexicans there is a 'new Berlin Wall' exemplified in the refortification of the fence along the US–Mexican frontier (see Nevins 2002 and Smith 1998).

Underlying these kinds of tensions and counterpoints, one can discern a deeply-rooted unevenness which is symptomatic of the process of globalization. Such an unevenness is tellingly depicted by the Cuban artist and writer Mosquera (1994), who argues that while the word 'globalization' may evoke the idea of a planet in which all points happen to be interconnected in a web-like network, in actual fact, connections occur inside a radial and hegemonic pattern around the centres of power, while the peripheral countries tend to remain disconnected from one another, or are only connected indirectly via and under the control of the centres. For Mosquera, there is a twin structure of 'axial globalization' and 'zones of silence' which forms the basis of the economic, political and cultural network that moulds the whole planet. In the highly centralized system of museums, galleries, collectors and market networks, Mosquera argues that the countries which host the art of other cultures are at the same time curating the shows, so that the world is being practically divided between 'curating cultures and curated cultures'.[1]

Mosquera's couplet of 'axial globalization' and 'zones of silence' highlights a key dimension of the geopolitical unevenness of globality, while also foregrounding the place of the periphery in contemporary treatments of culture. A primary theme here concerns the issue of how genuinely global is the contemporary theorization of global politics. In some instances, it is clear that a North–South divide emerges when the question of global change is posed. Nakarada (1994), for example, reporting on a workshop held in Zimbabwe, where the theme was the future of 'world order', noted the existence of a crucial North–South difference in the orientation of the discussion. The participants from the North tended to stress the phenomena of speed and the dissolution of spatial borders, with some emphasizing the positive potential of globalization. By contrast, participants from the South were far more negative in their diagnosis of globalization, referring to the South as a new object of recolonization and global apartheid.

This kind of split raises the question of the existence of a North–South divide in terms of the effects of globalization and also of the presence of a North–South differential in the manner that this divide is diagnosed. A contemporary example of the former differential can be seen in the current political debate on US–Latin American relations.

If, for instance, we look at the impacts of NAFTA for Mexico, it can be noted that while Mexican trade with the US has skyrocketed, going from US$36 billion in 1993 to $450 billion in 2002, most of the high-volume South–North exchanges have been between a handful of transnational subsidiaries in Mexico and their US-based corporate headquarters (*Latin America Press* no. 1, 15 Jan. 2003: 6). During these years, a substantial number of Mexico's banks have now become controlled by US, Canadian and Spanish investors, while US-based Wal-Mart has become Mexico's leading retailer. Mexican agriculture has been badly affected by imports – 6 million tons of cheap US and Canadian corn, much of it genetically modified, enters Mexico each year, displacing small farmers from the internal market. Moreover, farmers have now to sell to transnationals such as Cargill which have taken over the nation's privatized grain distribution infrastructure. Basic food imports have increased by 77 per cent over the past decade to an estimated $78 billion, equivalent to the government's public debt, and critics fear Mexico is losing its food sovereignty. There is inequality also in food subsidies, so while per capita subsidies of $21,000 per annum enable US farmers to sell produce in Mexico at prices 20 per cent below production costs, Mexican farm subsidies have continually fallen so that they are now at an average of $760 per farmer (*Latin America Press* ibid.).

With increased poverty in the rural areas of Mexico, many farmers have joined the immigration flow to the US. Further, in accordance with a 10-year-old NAFTA schedule, the programmed suspension at the beginning of January 2003 of all tariffs on basic agricultural imports from the US has provoked Mexican peasant farmers into organizing militant protests. During a recent demonstration (December 2002) of rural workers outside the Mexican Congress, where using sledgehammers and tractors as battering rams, the *campesinos* broke down the gates of the building, one 80-year-old peasant farmer from Guanajuato said, 'I'm an old man and I've never had to work in *El Norte* [the US] because my land gave me what I needed to live –but now this government is forcing me to go there' (*Latin America Press* ibid.).

The radical opposition expressed by Mexican peasant farmers to the subordinating effects of 'free trade doctrine' has been shared more generally. In late October 2002, in Quito, Ecuador, for example, at a meeting of trade ministers for the FTAA (Free Trade Area for the Americas), demonstrators who were organizing protests against the new trade deal for 2005 proclaimed 'We don't want to be an American colony!', linking their protests to the Brazilian President Luis Inácio Lula da Silva's description of the FTAA as a policy of 'annexation, not integration' (*Latin America Press* no. 24, 2 Dec. 2002: 4–5).

Similarly critical while broader evaluations of neo-liberal globalization were voiced in Havana in April 2000 when members of the Group of 77 came together to discuss North–South issues (see *Third World Resurgence*, no. 117, May 2000),[2] and such critiques continue to spread, as exemplified by Vindana Shiva's (2003: 87–8) statement that 'globalization is a project of domination by the North over the South, by corporations over citizens, by patriarchal structures over women, by humans over other species'. Overall, such representations, especially in relation to the terms of transnational integration, connect historically to José Martí's views on American economic integration in the late nineteenth century (mentioned in chapter 2), and raise the issue of the political nature of integration, of the balance between interdependence and dependence/domination. This is not a new theme.

Let us recall the oft-quoted passages from Marx and Engels' *Communist Manifesto*, passages which are used to support the view that in the mid-nineteenth century Marx and Engels already had a penetrating grasp of what is now called 'globalization'. For them, the bourgeoisie could not exist without revolutionizing the instruments of production, the relations of production and with them the entire relations of society. The bourgeoisie, through its exploitation of the world market, was portrayed as giving a cosmopolitan character to production and consumption in every country, and the introduction of new industries was seen as a 'life and death question for all civilized nations'; these new industries draw 'raw material from the remotest zones', and their products are consumed 'not only at home, but in every quarter of the globe', bringing about 'the universal *interdependence* of nations' and drawing 'even the most barbarian nations into civilization' (Marx & Engels 1998: 39, emphasis added). This notion of interdependence is almost immediately followed by a passage which highlights how the bourgeoisie has made 'barbarian and semi-barbarian countries *dependent* on the civilized ones, nations of peasants on nations of bourgeois, the East on the West' (Marx & Engels 1998: 40, emphasis added).

The annihilation of space by time, the increased speed involved in the formation of a world market, and the spreading out and modernization of the means of transport and communication, are linked to the interdependence of nations in general and the dependence of the non-Western on the Western nations. This duality finds an echo in the contemporary discussion of globalization, since the interdependence that Marx and Engels referred to is frequently associated with the Western world while the dependence of the non-West on the West represents that other aspect of global politics that is sometimes left in an investigative shadow. When this aspect is brought into the analytical arena, it is not infrequently connected to questions concerning the coloniality of power, Empire and imperialism, and to answering such questions as 'is globalization equal to Americanization?' (Bourdieu & Wacquant 1999). Alternatively, it has been argued that globalization is becoming increasingly decentred, and that for Giddens (1999: 16–17), for example, 'reverse colonization', where non-Western countries influence developments in the West, is becoming increasingly common, being exemplified through the latinizing of Los Angeles, the emergence of a globally oriented high-tech sector in India, and the selling of Brazilian television programmes to Portugal.

These issues are crucial to the analysis of geopolitical change in global times and link into problems raised in an emerging literature on 'post-modern geopolitics'. One of the key themes here concerns the predominance of flows over fixities, and the decreasing territorial power of nation-states. Campbell (1996: 18–19), for example, stresses the importance, after Foucault, of envisaging power as something which circulates, so that our post-modern era can be better formulated in terms of a 'centrifugation of power'. The analytical challenge is therefore to articulate an understanding of world politics that is attuned to the need to move beyond the sovereignty problematic, and in contrast appreciate the significance of flows, networks, webs and identity formations (Campbell 1996). A similar position is taken by Dalby (1999) and Ó'Tuathail (1998 and 2000), whereby the deterritorialization of risks and threats, informational capitalism, and the increased prominence of cybernetic spatiality are posited to be significant elements of a post-modern geopolitical condition. These are important points, but if we take a post-colonial perspective, the analytical landscape can be seen in a related but quite different light.

In moving the argument forward, I intend to examine two related themes: first, mutations in the imperiality of power, including questions of state sovereignty and the notion of a 'new American Empire', and second, the associated modes of representation of global politics in a frame of North–South relations. These themes form a key part of a post-colonial geopolitics in which the sovereignty problematic is still central and in which imperial power assumes a centripetal role.

The Colonial/Imperial Interface

It is sometimes the case that a discussion of definitions can help clarify the analytical route. Here, the context is formed by terms such as colonialism, imperialism and Empire. In relation to colonialism and imperialism, Said (1993: 8) defined imperialism as 'the practice, the theory and the attitudes of a dominating metropolitan centre ruling a distant territory'. In contrast, colonialism, which, for Said, was seen as being almost always a consequence of imperialism, was defined as 'the implanting of settlements on distant territory'. This implantation of settlements was of course only the most visible expression of an invasive and multi-dimensional power that Aimé Césaire (2000: 43) described in terms of 'cultures trampled underfoot, institutions undermined, land confiscated, . . . and extraordinary *possibilities* wiped out' (emphasis in the original).

While by the 1990s colonialism had been effectively brought to a close, having been declared illegitimate by the United Nations in 1960, and condemned as 'alien subjugation, domination and exploitation' and 'a denial of fundamental human rights',[3] imperialism would seem to be more enduring. Imperial politics can be linked to an invocation of a posited Western moral superiority and duty. In the West there has always been and remains the narcissistic assumption that the non-West could only be improved by becoming more like the West itself. In this context, therefore, the geopolitics of representation of the other has been and remains a crucial mechanism of imperial power, a point to which I shall return below.

In order to pursue the further delineation of colonialism from imperialism the following three points can be made.

First, the term 'colonial' is frequently used metaphorically, as in the example of the Quito protests against further neo-liberal integration in the Americas, which was likened to a new kind of colonialism. This is a widespread contemporary tendency, so that, for example, the recent Plan Puebla-Panama, initiated by the Mexican president Vicente Fox in 2001, has been contextualized as a process of the 'recolonization' of the indigenous peoples of southern Mexico and Central America (Moro 2002). In an African context, a Ugandan political scientist describes global politics in the mid-1990s as being primarily characterized by a 'recolonization of subject peoples' (Tandon 1994), and as a final example, a resurgent US imperialism has been linked to a new 'global colonialism' (González-Casanova 1995) and a new 'recolonization of the world' (Maillard 2003). These examples reflect to some extent the comparative facility of the term in contrast to say the 're-imperialization of the world', but they also signify an attempt to give adequate expression to the strength of opposition to the resurgence of invasive Western power.

Second, there is another approach to colonialism, or what is referred to as the 'coloniality of power', that originates in Latin America with the work of the Peruvian sociologist Quijano (2000) and the cultural studies and literary theorist Mignolo (2000b and 2000c). According to Quijano, what is referred to as globalization is in fact the culmination of a process that was initiated with the formation of a colonial/modern Euro-centred capitalism as a new global power. One of the crucial components of this model of power is the social classification of the world's population around the idea of race, a mental construction that expresses the basic experience of colonial domination and one that still pervades contemporary global power. For Quijano (2000: 533) the racial axis has a colonial origin, but it has proved to be more durable than the colonialism in which it was established, so that the 'model of power that is globally hegemonic today presupposes an element of coloniality'. The advantage of this kind of perspective lies in the emphasis it gives to the historical origin and continuity of colonial power and the relation of this kind of power to questions of race and cultural difference. However, it also leaves open the question of how one might distinguish the coloniality from the imperiality of power in which the actual possession of colonies is no longer a necessary condition for maintaining imperial control.

Third, there is the question of how the term 'Empire' is used and how this term connects with imperialism. Doyle (1986) offers us a strict definition of Empire as embracing a formal or informal relationship in which one state controls the effective political sovereignty of another. This can be achieved by force, by political collaboration and by economic, social or cultural dependence. Imperialism can then be seen as the process or policy of establishing or maintaining an Empire. With the use of the term Empire, one of the main points of contention concerns the geopolitical scope of this kind of power, since in the case of the US, for example, its relations with Western Europe and Japan are qualitatively different from its relations of power with societies of the South (McGrew 1994). Furthermore, there are other writers who strongly oppose the use of the term Empire in relation to the US, asserting that

imperial power is sovereign power and sovereignty means a direct monopoly control over the organization and use of armed might, direct
control over the administration of justice and the terms of trade. Since
the US does not have anything like such direct authority over other
countries, notions of an 'American Empire' are misconceived (see Zelikow 2003). Along a different track, Hardt and Negri (2000), as we have
already seen, posit that the US occupies a privileged, but not necessarily
determining, position in a contemporary Empire, or 'post-modern imperialism', that is rooted in the world market and diffused through
networks of changing coalitions and associations.

Clearly, this overall theme, and especially the last point concerning
Empire, require further elaboration. This will be done through examining the themes of state sovereignty, interventionism and the resurgence of
imperial power.

State Sovereignties and the Persistence of North/South Differences

The sovereignty of states has traditionally been defined in terms of the
exercise of a legitimate and supreme power within strictly delimited
territorial boundaries. Sovereignty, from the inside, can be envisaged in
relation to three concerns: the maintenance of the geopolitical frontier of
a society; the construction of a legal and institutional system; and the
relations between state and civil society, including the preservation of
internal order. In the international realm, in a world of supposedly
autonomous states engaged in an unregulated contest of wills, sovereignty from the outside is both a building block of the international
system and a problem to be overcome (Walker 1993). While it can be
justifiably argued that such an inside/outside distinction is relevant, it
needs to be supplemented by an appreciation of a significant North/
South differential.

This overall differential is rooted in a contextualization of sovereignty
and self-determination that has been historically affected by the dependence of international law on Western norms and the intimate association
between the rules, practices and processes of international law and
politics and the development of the West's imperial power (Grovogui
1996). In particular, from 1492 onward, the erasure of the fact of
conquest from the discourses of law and international politics enabled
Western powers to claim rights and privileges that would otherwise have

been considered illegitimate. As Strang (1996: 43) has put it, 'the imperial moment took place within and was carried forward by a collective delegitimation of the sovereignty of non-Western peoples'. Subsequently, with the twentieth-century emergence of newly independent states in Africa and Asia, that followed on from the earlier nineteenth-century lead of independent Latin American states, the significance of state sovereignty was strongly asserted. This was reflected in the Charter of the Organization of African Unity adopted in 1963, and also in the Charter of the Economic Rights and Duties (CERDS) adopted by the UN General Assembly in 1974, which included a call for the global economy to be regulated by states (see Clapham 1999).

Although the states of the South have given key importance to their own sovereignty, there has been a tendency in the North to regard the sovereignty of states of the South as being characterized by a lack of effectiveness and modern authority, so that terms such as 'quasi-states' are sometimes employed to situate the state sovereignty of Third World countries.[4] This tendency has become more overt, for instance 'quasi-states' becoming 'failed states' (Gros 1996 and Huntington 1998), as the following phenomena have become more visible and often more acute: the disintegration of centralized state authority as in countries such as Somalia and Sierra Leone; the reduction of the territorial extension of state power through the existence of effective and durable guerrilla organizations, as is the case in Colombia and also in Sri Lanka; and the territorial fracturing of the sovereign power of the state, as can be seen in the cases of Angola and Afghanistan, due to long periods of civil war and inter-regional strife. Moreover, with the increase in wars, ethnic conflicts and humanitarian crises in the post-Cold War period, sovereignty has come to be seen in a dual sense.

For Kofi Annan (1999), for example, there are two concepts of sovereignty, one based in the UN Charter which guarantees state sovereignty and its territorial integrity, and a second originating in the Universal Declaration of Human Rights (UDHR) of 1948 whereby individual human rights are seen as central. For Annan there is nothing in the UN Charter which precludes a recognition that there are rights beyond borders, but there is ambiguity in the sense that the charter says that armed force shall not be used, 'save in the common interest', without specifying what is the common interest and who should define it. There is a clear inconsistency in international law in the sense that the UN Charter guarantees the rights of states, whereas the Universal Declaration of Human Rights guarantees the rights of individuals that might well be violated by the actions of states. In the more recent context of 'humanitarian interventions', as in Bosnia and Kosovo, the UDHR has been regarded as an essential basis of legitimization.

Kofi Annan goes on to suggest that there are four aspects of humanitarian intervention that are worthy of consideration: (a) that a new commitment to humanitarian intervention must be seen to be a universal commitment; (b) that more thought needs to be given to how 'national interest' can be dove-tailed with a 'collective interest', i.e. the interest of humanity at large; (c)that where forceful intervention becomes necessary, as for example in the case of genocide in Rwanda, the Security Council must be able to rise to the challenge as the defender of common humanity; and (d) that the international commitment to peace must be as strong as the commitment to war – in other words, for the Secretary General, the securing of peace and stability in a case of humanitarian tragedy must be the paramount objective of intervention.

In this context, advocating the transgression of state sovereignty can be justified by the existence of a severe humanitarian crisis. In addition, it is argued that if states bent on 'criminal behaviour' realize that 'frontiers are not an absolute defence', such behaviour will be less likely to re-occur in the future and the common interest of humanity will be furthered. I have given some attention to these points because, even though my own analytical context is not specifically one of humanitarian interventionism, what emerges from this short article can be located in a much wider framework that is relevant to my own narrative.

First, one needs to question who it is that defines what constitutes the 'universal', and whether, for example, such a specification should rest with the permanent members of the Security Council, or with the General Assembly, or with say the G-8 or the G-77? There is here, as we have seen previously, a crucial issue concerning the cultural politics of translation. Any concept of universality cannot cross the cultural borders it professes to cross without an act of translation – unless, that is, universality is only associated with an expansionist Western logic (Butler 2000).

Second, the phrase 'national interest' is most appropriately understood in accordance with the *unevenness* of the state powers that defend their own 'national interests' in contradistinction to any conception of a 'collective interest' of humanity at large, although as we shall see, 'national interest' can be represented as being equivalent to a wider global or collective interest.

Third, in a world of globalization and interdependence, one might be reminded of the duality in Marx and Engels' 1848 *Manifesto* whereby capitalist-led interdependence sits uneasily with the reproduction of dependent relations between the so-called 'civilized' and 'barbarian' or 'semi-barbarian' nations. What does today's 'actually existing' interdependence of nation-states look like? Do we inhabit a globalized world in which the territorial differences between nation-states are only differences of degree, and how interdependent are the states of the South with the states of the North in a globalized world?

Fourth, commitment to peace and 'nation-building' in the wild, border zones or 'global frontier-land' of the Third World poses the question of whether such Western-controlled nation-building can ever be free from the 'national interest' of those same Western nations. Is 'temporary imperialism', or 'Empire Lite', as Ignatieff (2003: vii) recently contends, the 'necessary condition for democracy in countries torn apart by civil war'?

The above issues concerning 'national interest', '(inter-)dependence' and 'universality' can be usefully linked to sovereignty, powers of intervention and the difference that North–South relations make. A key question with all these issues relates not only to who has the power to define sovereignty and to effect geopolitical interventions, but also how are these interventions legitimized? It is here again that the imbrication of power and discourse is so crucial, within which the interconnections between the desire, will and capacity to act and the legitimizing regimes of representation are of primary significance.

As a symptomatic illustration of these kinds of interconnections, it can be suggested that international, and more specifically North/South, relations are characterized by an asymmetrical power to define and be defined – who, for example, has the will and capacity to define sovereignty for both the self and the other? Looking at the example of the US in relation to societies of the Latin South, this asymmetrical power, notwithstanding its varied mutations, has been continuously present, as has been discussed in previous chapters.

Such an asymmetrical power has gone together with a sharp dissonance between the imperialist impact of the US and its enunciated belief from Jefferson on in the self-determination of peoples. This contradiction has been the subject of varied attempts at resolution. In some instances, as with the Reagan administration's intervention in Grenada in 1983, a clear separation was drawn between a people that purportedly needed rescuing – the US invasion of Grenada was portrayed as a 'rescue mission' – and a tyrannical regime that not only did not represent the interests of its people but also constituted a threat to the neighbourhood of the Americas (Weber 1999: 79). At the same time, this separation has been reinforced by the association of the 'tyrannical regime' with a foreign ideology that is seen as subverting the values of Western freedom and democracy, as in the Cuban case both during the Cold War and also post-1989.

In these contexts, the US has assigned to itself the right and also the responsibility to define the sovereignty of another people. For example, in the case of the US invasion of Panama in December 1989, which was undertaken despite the refusal to accord it legitimacy at either the continental, i.e. OAS (Organization of American States), or the UN level, the US Permanent Representative to the United Nations declared that 'the sovereign will of the Panamanian people is what we are here defending' (qtd in Weber 1995: 100). The hegemonic will to define the sovereignty of another people is also clearly present in the Cuban case, and not only in relation to the Platt Amendment, as discussed in chapter 2, but more recently in the example of the Cuban Liberty and Democratic Solidarity Act of 1996 (US GPO 1996), sometimes known as the Helms–Burton Act.

This Act contains a number of statements that establish a clear demarcation between the rights of the Cuban people and the Castro government. It is written, for example, that it has been the consistent policy of US administrations to 'keep faith with the people of Cuba', while sanctioning the 'totalitarian Castro regime'. The Cuban people, the Act continues, 'deserve to be assisted in a decisive manner to end the tyranny that has oppressed them for 36 years', and the 'continued failure to do so constitutes ethically improper conduct by the international community' (US GPO 1996, 110 Stat, 786–8). The Act set out a number of purposes which can be summarized as follows.

First, the Act aimed at encouraging the holding of free and fair democratic elections in Cuba under international supervision, and secondly, it was stipulated that international sanctions against the Castro government should be strengthened. Third, the Act aimed to protect US nationals against confiscatory takings and the exchange of confiscated property, and fourth the Act had the purpose of providing for the continued national security of the US in the face of what were considered to be continuing threats from the Castro government. In addition, the Act provided a 'policy framework for US support to the Cuban people' in their need for a transition to democracy. This framework followed on from the Cuban Democracy Act of 1992, that encouraged the governments of countries trading with Cuba to restrict their trade and credit relations with the island. Overall, the 'Cuban problem' was constructed as both a domestic and international issue.

Thus, while a specific conception of the Cuban people, with an emphasis on their right to freedom, democracy and prosperity through 'self-determination' (US GPO 1996, 110 Stat, 789 and 805) is set against a notion of a tyrannical regime that needs to be replaced, that same tyrannical regime is constructed as a threat not only to its own people but also to the US. Its removal therefore is interpreted as being beneficial to both the Cuban people and the people of the US and its government. Symptomatically, while 'regime change' is the clear long-term goal, at the same time, the Act underlines the fact that US policy is to (a) 'recognize that the self-determination of the Cuban people is a sovereign and national right of the citizens of Cuba which must be exercised free of interference by the government of any other country' and to (b) 'encourage the Cuban people to empower themselves with a government which reflects the self-determination of the Cuban people' (US GPO 1996, 110 Stat, 805). Reminiscent of the third article of the Platt Amendment, whereby the Cuban government had agreed to the right of the US to intervene in order to preserve Cuban independence, the US government in 1996 defined Cuban sovereignty as a Cuban right which must be exercised free of interference by the government of any other country. At the same moment, the US was itself setting the terms for the nature of that sovereignty and stipulating its own right and responsibility (as well as that of the international community) to act against the present Cuban government. And this government was represented as being separate from its people, as implicitly being an extraneous and unhealthy growth on the body politic of the Cuban people.[5]

From the Cuban Liberty and Democratic Solidarity Act of 1996 it can be seen that the US has assumed the right to 'represent' or act as 'guardian' for the Cuban people – to write a narrative for the defence of Cuban sovereignty which is also a defence of US interests against an ostensibly threatening and tyrannical regime. The internal and external are closely intertwined, and the moral, cultural and political leadership role assumed by the US can only be fully understood as part of the geopolitical history of US–Cuban relations (see, for example, Benjamin 1990 and Weldes & Saco 1996). Equally, however, as writers such as Cottam (1994) and Kenworthy (1995) have demonstrated, US hegemony has been constructed in similar ways with regards to other Latin American societies, where the combined US role of being 'teacher', 'doctor' and 'policeman' has received a continuing significance (Schoultz

1999). But also, the invocation of a commonality of interest must not be forgotten. This is illustrated, for example, in the emphasis in the Helms–Burton Act on the posited shared US–Cuban interest in representative democracy, market economy and freedom and prosperity. Moreover, there is the shared importance given to self-determination, which is born out of a sense of the American nations being the product, as President Kennedy expressed it in 1961, of a 'common struggle, the revolt from colonial rule' (see Holden & Zolov 2000: 227).

From the perspective of Washington, what has been specific about contemporary Cuba, or more accurately the Castro government, is its alliance during the Cold War period with the Soviet Union and its continuing adherence post-1989 to a communist political system. This specificity has been used by the US as a justification for not only classifying Cuba as a threat to the US and the Western Hemisphere, and now more recently as a 'rogue state', but also as a justification for its embargo on or blockade of Cuba – an embargo which has been condemned by the UN, the European Union and the Inter-American Juridical Committee, which has ruled that such a measure against Cuba violates international law (Chomsky 2000: 2). The embargo and related acts of interference[6] also contravene Article 15 of the 1948 Charter of the Organization of American States, which stipulates that no state or group of states has the right to intervene, directly or indirectly, in the internal or external affairs of any other state (see Holden & Zolov 2000: 192).

The Cuban case might be considered to be too specific to act as a basis for broader interpretations, and certainly in the context of US–Latin American relations it has a unique geopolitical significance. However, it can also be argued that US–Cuban relations express in a rather concentrated form a persistent theme in the history of US–Latin American relations. For example, there are other cases of the US assigning to itself a role of framing relations with Latin American societies in which their sovereignty is transgressed. One thematic exemplification of this problem concerns the US policy of 'certification' in relation to the 'war on drugs'.

The US certification process was initiated by President Reagan in 1986 as part of Washington's 'war on drugs' strategy, and has become the essential point of US influence over other countries' drug control policies, with an international link being made to the 1988 UN Vienna Convention Against Illicit Traffic in Narcotic and Psychotropic Drugs. Every February, the White House announces the results of the certification process, and countries regarded as major drug producing or transit countries are examined on their judged efficiency in drug control during the previous year. If their efforts are deemed to have been unsatisfactory, the offending countries are 'decertified'. Decertification of a country can lead to the imposition of mandatory sanctions, including the suspension of at least 50 per cent of US assistance for the current year (excepting humanitarian aid and drug cooperation aid), a complete suspension of aid for the following years and a required US vote against loans for the decertified country. Further discretionary sanctions can include the denial of preferential tariff treatment under existing acts, the curtailment of air transportation and traffic between the US and the country in question, and the withholding of tourist visas to the US.

This certification process is undertaken for a whole range of countries of the South, not just for Latin America, and this is a clear case of the US assigning to itself the authority to determine not only whether other countries are complying with US stipulations but also to evaluate unilaterally their compliance with the UN Vienna Convention. In 1997, Mexico was threatened with decertification, and the discussions that took place in both the US and Mexican Congresses were particularly revealing.

In Congressional hearings on Mexico's certification, Mexico was consistently described as being rampantly corrupt and the source of most of the drugs coming into the US. The US was said to have a duty and a right to correct Mexican behaviour. For example, Senator Trent Lott, speaking before the Senate on 20 March 1997, stated that 'we have a right' to have Mexican nationals extradited to face charges in the US and to have DEA (US Drug Enforcement Agency) agents carry weapons in Mexico. The Clinton administration was given 90 days to provide evidence that Mexico had made substantial progress in cooperating with US drug control efforts, including demands that Mexico allow more US law enforcement agents into Mexico, permit agents to carry weapons, extradite Mexicans to the US and permit the US Coast Guard to chase traffickers and stop them in Mexican waters. If Mexico did not agree to these measures decertification would follow (Cottam & Marenin 1999: 225).

Mexican responses to the US perspective were fast and furious. The Mexican Congress voted unanimously to condemn certification in principle as an insult to national sovereignty. The US was also condemned for its hypocrisy on drug production, it being noted that most of the marijuana consumed in the US is grown in the US. Overall, as Cottam and Marenin (1999: 223) explain, the Mexican government has not only regarded the certification process as a violation of international law but also as an illustration of US arrogance and imperialism.[7]

The assertiveness of US officials has also been expressed with regard to other Latin American countries. In one instance, as Der Derian (1992: 108–9) informs us, the perceived 'narco-terrorist' threat from Colombia provoked Mayor Koch of New York to advocate an air attack on Medellín, while Daryl Gates – the chief of the Los Angeles Police – went further and called for an outright invasion of Colombia. Nor have such views been necessarily exceptional, as Gamarra (1999: 198–9) shows in the case of the war on drugs and US–Bolivian relations in the late 1990s.

The overall development of a US strategy on drugs has created a concentration on what has been referred to as the 'supply side' of the problem. Hence, during the 1990s US strategy was driven by the desire to destroy or at least drastically curtail the cultivation of coca leaves from which cocaine has been produced and distributed throughout the US and beyond. To contain the problem at source, thus legitimizing a new round of interventions, has been a guiding priority of the Washington strategy, and most acutely in the Bolivian case, a US military presence has raised important questions of sovereignty and national autonomy (Gamarra 1994 and 1999). In the Bolivian example the drugs problematic has been associated with a double transgression of sovereignty. So, not only has the certification process been regarded as an example of the US unilaterally monitoring and judging Bolivia's national policy on a key question of commerce and security, but also the presence on Bolivian soil of US military advisers since the early 1990s has been viewed as a further infringement of Bolivia's territorial sovereignty. In contrast, from the perspective of Washington, the certification process has the essential function of acting as a monitoring and evaluation mechanism for the countries of the South, being related to the earlier Foreign Assistance Act of 1961, which is referred to in President Clinton's 1997 memo on 'drug war certification' (White House 1997).

The certification process has been and continues to be used as a form of leverage over and/or as an expression of disapproval of Third World societies. This is clearly evident in the case of Iran, which despite being a country with draconian anti-drug policies that include the massive forced eradication of illicit crops and the execution of drug traffickers, is annually decertified by Washington (Joyce 1999: 212). The certification process is an example of an asymmetrical form of power which once stimulated the former Brazilian president Fernando Henrique Cardoso to sardonically remark that Latin American countries might well think of periodically publishing reports on the US (*Actualidad Latinoamericana*

2000: 15). This kind of inversion is also relevant for other issues, including, for example, the US record on human rights (Chomsky 2003).

In the above section I have looked at two issues relating to sovereignty and the persistence of North–South differences, and this has been done in the context of the will and capacity of the US to extend its power of monitoring and intervention well beyond its own borders. It has to be borne in mind that the examples I have taken are only illustrative of the limiting of sovereign powers in the North–South context, and there are other ways of treating sovereignty questions in such a context (see e.g. Nordstrom 2000 and Ong 2000). In my own approach, I have been more concerned with the impact of the US on the meaning and enframing of sovereign powers. Thus, in the case of Panama, and especially Cuba (and these two cases are not isolated examples), it is clear that the US gives itself the role of defining and representing the sovereignty of other peoples against their own governments which in specific periods are considered to be tyrannical and illegitimate.[8] In the case of drug trafficking, the US power to certify or decertify other countries constitutes another key example of a 'power-over' – a form of hegemonic authority, which contrasts with the notion of a globalizing world that is rooted in interdependent relations. These are only two examples which relate directly to the US role in world affairs; and as we have seen in chapter 4, there are other examples under neo-liberal globalization of the financial and economic sovereignty of Third World countries being transgressed through the operations of international financial institutions such as the World Bank and the International Monetary Fund. My point here has been to concentrate specifically on a certain kind of imperial power which is specific to the US. My perspective needs to be further developed in the context of the geopolitical world, post-9/11.

Tracing the Resurgence of Imperial Power

At the beginning of the twenty-first century, the resurgence of US power has been contextualized in terms of the disintegration of the Soviet Union, the projection of a neo-liberal globalization, based on a US version of 'open economies' and 'streamlined states', a unique military prowess and the absence of any rival 'multi-dimensional power' (Cox 2001). More specifically, Nye (2002), for example, argues that the US wields three kinds of power: (a) military, where the US is the only country with both nuclear weapons and conventional forces with global reach, and where US military expenditures are greater than those of the next 8 countries combined; (b) economic, where the US share of world production is equal to the next 4 countries combined, a datum to which might be added the fact that in the mid-1990s around 26 per cent of total world foreign direct investment originated from the US, more than twice as big a share as the next largest source countries, Japan and the UK (Dicken 1999: 36); and finally (c) cultural, whereby the US is far and away the number one film and television exporter in the world, while its colleges and universities act as a magnet for foreign students. It is in this kind of context that Nye suggests that although globalization is not just Americanization, it is 'America-centric', since so much of the information revolution comes from the US and a large part of the content of global information networks is currently created in the US. For Nye, its cultural projection in the world reflects its unique 'soft power', which is paradoxically vulnerable to an over-assertive projection of 'hard power', a military dominance that is expressed unilaterally. In an interconnected, globalized world, Nye argues that the US, despite its pre-eminent global power, needs to be multilateral to achieve its objectives.

Nye's notion of 'multilateralism' as an effective strategy for the securing of Washington's global power rests on a notion of hegemony linked to leadership, persuasion and the engineering of consent. It is tied into a conception of globalization which sees the US as the leading power acting for a greater good. It is the 'indispensable nation', or the 'only comprehensive global superpower' (Brzezinski 1997: 24), that seeks to enforce the global rules. President Clinton expressed the point by underlining the fact that the US had a key role in 'helping to write the international rules of the road for the twenty-first century'. For Clinton, speaking in 1999, in the context of the Kosovan conflict, the world was witnessing a 'great battle between the forces of integration and the forces of disintegration; the forces of globalism versus tribalism' (qtd in Bacevich 1999: 10–11).

The US role of enforcement, of the deployment of military power, is an important expression of the imperiality of power, ever more evident with the invasion of Iraq in March 2003. Nevertheless, it ought not to distract us from the pivotal significance of the vision of 'global America' and its mission in the world which is constructed as a legitimizing basis for the overall exercise of power. This theme has been examined for earlier periods (see chapters 2, 3 and 4 above), and for the post-Cold War era concepts of democracy, freedom and prosperity, and market economics, US-style, have provided the solid basis for justifying the export of the

'American way of life'. Continuing the same theme, the Clinton years were characterized by a strategy of 'democratic enlargement' and free trade.[9] Although the 1990s saw the identification of 'rogue states' as a new threat to the US and the West, there was no explicit objective of 'regime change' nor of a doctrine of 'pre-emptive strikes' against such states; rather, a targeted strategy of containment was deployed against states such as Iraq, Iran and North Korea. Conversely, the Bush strategy from 2000 onwards has expressed a more systematically aggressive approach to global politics in general, linking back to the Reagan years of the 1980s.

Democracy, however, remains a central organizing principle. For example, Wolfowitz (2000: 39–40), at time of writing the Deputy Secretary of Defence in the Bush administration, looking back on the Cold War period, stresses the importance of the link between promoting democracy abroad and the advancement of American interests. This spread of the 'democratic principle' has to be part of a strategy of building successful coalitions, so that the US exhibits the importance of world leadership, 'demonstrating that your friends will be protected and taken care of, that your enemies will be punished, and that those who refuse to support you will live to regret having done so' (Wolfowitz 2000: 41). Punishing enemies is clearly the focus of 'regime change', and while Kagan and Kristol (2000), directors of the 'Project for the New American Century', also underscore the importance to the US of promoting liberal democratic governance throughout the world, they also emphasize the need to change regimes that are judged to be tyrannical. The ultimate goal with states like Iraq, Serbia and North Korea should be 'transformation' rather than co-existence. Their vision is traced back to Theodore Roosevelt's concept of a robust brand of US-led internationalism. For Kagan and Kristol (2000: 68–9), Roosevelt rightly insisted that the defenders of civilization must exercise their power against civilization's opponents – quoting Roosevelt directly, they concur with his statement that 'warlike intervention by the civilized powers would contribute directly to the peace of the world'.

'Liberty' and 'democracy', 'civilization', 'American internationalism', a doctrine of 'pre-emptive strikes' against rogue states leading to 'regime change', are all markers that are found in President George W. Bush's *The National Security Strategy of the United States of America*, published in September 2002 (White House 2002). The context of this orientating document is mainly formed by the events of September 11, but there are also themes which express a longer provenance. When President Bush, for example, asserts that the US has long maintained the option of pre-emptive actions to counter a sufficient threat to our security, he is drawing an implicit link with the actions of the Reagan administration, and specifically the bombing in 1986 of Libya, which was justified as 'self-defense against future attack' (Chomsky 2003: 68).

What is specific to the Bush Doctrine is the linking of an imminent threat to the US from 'rogue states' and terrorists with their posited ability to use weapons of mass destruction. This is then associated with the need to defend civilization, and if necessary by acting pre-emptively (White House 2002: 15). There is a clear identification of threats to the US's and the civilized world's security, especially from rogue states and terrorists, as well as from 'failed states' (ibid.: 1), but also (and crucially) there is an affirmation of 'America'. The US national security strategy, Bush states, will be based on a 'distinctly American internationalism' that will help to make the world not just 'safer but better' (ibid.). The US is defined as a 'great multi-ethnic democracy' that must provide a lifeline to 'lonely defenders of liberty' and 'look outward for possibilities to expand liberty' (ibid.: 3). Overall, this affirmation of 'America' is joined to a reassertion of the 'essential role of American military strength' (ibid.: 29).

Interspersed throughout the National Security Strategy document are references to the new dangers of living in a more interconnected world, and from the concluding section of the document, four points can be highlighted:

- the underscoring in global times of the increased intersections between the domestic and the foreign;
- the continuing need for the US to assume its role of global leadership;
- the desirability where possible to engineer coalitions of support for US policy (e.g. in the invasion of Afghanistan in October 2001);
- the will when necessary to act unilaterally (e.g. in the invasion of Iraq in March 2003, undertaken without UN approval).

The entwinement of the domestic and the foreign has been manifest in the reorganization of security at home with the establishment of a new Department of Homeland Security and the elaboration of an aggressive strategy of pre-emptive war abroad. The direct defence of the US homeland has been designated 'Operation Noble Eagle' and comprises, among other things, actions to protect civil population centres and critical infrastructure. 'Offensive operations', as the Chairman of the Joint Chiefs of Staff, General Myers, defined it, are labelled 'Operation

Enduring Freedom', and during the attack on Afghanistan included a combination of ground, air and naval operations, wherein 'versatility' and 'flexibility' were considered crucial to the success of the war. 'Consider the examples,' General Myers comments, 'of forward air controllers on horseback and special operations troops transporting their high-tech gear on donkeys to isolated mountain tops from which they directed strikes of precision guided munitions' (Myers 2002).

On the domestic front, the passing into law of the Patriot Act in October 2001, and more recently the Domestic Security Enhancement Act, or Patriot Act II, in early 2003, reflect not only an increased concern with internal security but more significantly a willingness to undermine civil liberties in the cause of 'fighting terrorism' (Rivière 2003 and Williams 2003).[10] In addition, the US has been heavily criticized by Amnesty International for its inhumane treatment of prisoners, so-called 'unlawful combatants' held at Camp X-Ray on the US Naval Base at Guantánamo Bay, Cuba (see *The Independent*, London, 11 Jan. 2003: 13). On the foreign front, the US Central Intelligence Agency has again been granted the 'licence to kill' previously rescinded by President Carter in 1976, and as an example, in November 2002, six suspected Al Qaeda militants were 'taken out' or assassinated in the Yemen by a Predator missile fired from a US warplane in the Gulf (*El País*, Madrid, 9 Dec. 2002: 4).

Although it should be remembered that with the war in Afghanistan the US did succeed in assembling a coalition of support and secured UN approval for its actions, overall the US has become more markedly unilateralist in its foreign policy. Failure to ratify the Kyoto Protocol, the introduction of tariffs on steel imports, contradicting the free trade nostrum of the World Trade Organization, opposition to the International Criminal Court (White House 2002: 31), the invasion of a sovereign nation without the approval of the UN Security Council, and a state-sanctioned policy of selective assassinations, represent some examples of an imperial power acting in a unilateralist manner. As we saw in previous chapters, the unilateralist dimension of US foreign policy has a long history, as does the belief that the US has a right to act above the law when its own perceived interests are imperilled. The Bush administration is not the first US administration to behave in such a way, but its lack of interest in concealing such a strategy is quite distinctive. How does one set about explaining the expansionist and unilateralist nature of current policies, and what are the implications for the way we situate globalization in a post-colonial frame?

In developing an interpretation of the resurgence of US power I would argue that it is important to avoid uni-dimensional theories, especially those which centre on the economic – assuming, for instance, that the contradictions of capitalist accumulation produce certain political effects, and that there is a determining economic functionality to power. This does not mean that economic questions are not important – they certainly are, but they need to be integrated within a broader explanatory frame. Hence, as an example, Michael Klare (2001 and 2002) makes a powerful case for the key importance of the securing of vital resources for future US economic stability and growth. He suggests that in the 1990s the Pentagon began to give increased attention to regions such as the Persian Gulf, the Caspian Sea basin and the South China Sea, and that this shift in strategic geography reflected a new emphasis on the protection of vital resources, especially oil and natural gas. He cites a 1999 White House Annual Report on US security policy that stresses the strategic importance of ensuring access to foreign oil supplies, so that a main aim of policy ought to be that of providing stability and security in key producing areas so that US access to and free flow of these resources is ensured. Klare argues that resource wars are going to become the most distinctive feature of the global security environment. This he suggests is reflected in the following trends: the priority accorded to economic considerations by national leaders, the ever-growing demand for a wide range of basic commodities, looming shortages of key materials, social and political instability in areas containing major reserves of vital commodities, and disputes over the ownership of crucial resources.

Clearly Klare's argument can be used at least in part to explain US actions in Afghanistan, with its links north to oil-rich former Soviet states such as Uzbekistan, Tajikistan and Kyrgyzstan, and especially in Iraq with its extensive oil reserves. Is this then a sufficient reason to account for a new kind of US expansionism? It might be argued that another factor concerns the perceived decline in US power and influence during the 1990s and the need to counter such a posited decline.

Petras (2002), for example, points to such phenomena as: the failure of US policy to isolate the 'rogue states' of Iran and Iraq, the former successfully signing nuclear power agreements with Russia and oil contracts with Japan, and the running up of a US$430 billion trade deficit in 2000 while Western Europe's 350 million consumers were increasingly purchasing European-made goods. Furthermore, following Petras, in Latin America US hegemony is being severely tested by guerrilla movements in Colombia, by the populist reforms of Venezuela's President Chávez, and by a wide range of social movements in Ecuador, Brazil, Bolivia, Mexico and elsewhere. These trends provide an important contextual factor, just as also does the perception of a world characterized by increasing chaos, with 'failed states' and 'rogue states' constituting potential threats to the global reach of US power. Also, although Huntington's (1998) notion of a 'clash of civilizations' is implicitly repudiated in the 2002 National Security Strategy document, his dystopian vision of a global breakdown of law and order, with increasing anarchy in many parts of the world, has had an influence on the orientation of strategy, strengthening those voices who seek a stronger role for the military in US foreign affairs. Certainly, the idea of a world in increasing chaos lends itself to the call for a global enforcer to restore security and guarantee a new round of freedom and prosperity, and General Myers (2001) recently defined the Afghanistan operation in terms of the armed forces 'toiling on the fields of freedom and democracy'.

Another factor which leads on from the above, concerns the realization that for the first time in US history, the military dimension of power, the capacity, reach, technological advancement and speed of deployment of US armed forces, gives the imperial state an unparalleled edge over all its rivals. If US strategy is to guarantee the imperial state its long-term global pre-eminence, why not make more use of this exceptional war-making power to ensure that pre-eminence, through which its rivals can be disciplined and other recalcitrant nations can be subdued through 'shock and awe'. Again this might not be an entirely new perspective, as some similar arguments were aired in relation to the US potential to fight 'limited nuclear war' with the Soviet Union in the early 1980s, but today the preponderance of US military capability in a uni-polar world is something quite unprecedented.

These factors, the drive to control resources, a perception of a previous decline of US influence in a world of growing disorder, and the awareness of the existence of an overwhelming US military power, can all be referred to as contributing elements in an explanation of a 'new imperialism' or of a new strategy of aggressive expansion to secure US pre-eminence. But these factors do not of course necessarily of themselves lead to predictable outcomes in terms of strategy and action; an actual outcome or power-effect requires some kind of discursive mobilization produced by intervening agents of representation. In this case, the group organizing the 'Project for a New American Century' (PNAC) provides a significant example.

Established in 1997 as a non-profit, educational organization, the PNAC aims to promote US global leadership. In the founding statement of principles, it is asserted that during the Clinton administration foreign and defence policy went adrift, and what is required is a case for the re-assertion of 'American global leadership'. The history of the twentieth century, the statement goes on, shows that it is important to shape circumstances before crises emerge. The statement concludes by identifying four objectives: (a) to increase defence spending so as to carry out America's global responsibilities; (b) to strengthen US ties to democratic allies and to challenge regimes hostile to US interests and values; (c) to promote the cause of political and economic freedom abroad; and (d) to accept responsibility for America's unique role in preserving and extending an international order friendly to US security, prosperity and principles.[11] The signatories of this statement include Dick Cheney, Donald Rumsfeld and Paul Wolfowitz, all key figures in the Bush administration. The views held by these signatories to the PNAC statement are not unique, and in fact are reflected in influential think-tanks such as the American Enterprise Institute, the Cato Institute and the American Heritage Institute. The main point here is that the overall vision held by the supporters of the PNAC has become closely associated with the Bush administration's foreign policy and provides a particularly clear example of the significance of the impact on US imperial strategy of a well-organized 'pressure group' that at a particular moment has found access, through key appointments, to crucial levers of governmental power.

A final factor of crucial importance has been the impact of the events of 9/11. As a crystallizing, precipitating event, 11 September 2001 gave the supporters of the PNAC and other Republican groups a powerful opening to push through their four objectives mentioned above. Through the appointment of Dick Cheney as Vice-president, and his successful bringing into government of PNAC supporters, the PNAC has been able to use September 11 as an example of the need for a much more aggressive strategy of extending and intensifying US military power. The strategy is not without its contradictions and limits. Rising defence expenditures will have a detrimental effect on the US economy, and equally the accentuation of 'hard power' will certainly undermine the efficacy of the 'soft power', to use Nye's terms. The rise of opposition to US strategy and divisions between continental Europe and the US are already weakening the global hegemony of the US: remembering that that hegemony, to be effective, must include a capacity for consent and persuasion. Outright attempts at domination – a 'might is right' perspective – will tend to undermine the long-term hegemony of the US.

The Imperial, the Global and the Post-colonial

In bringing this chapter to a close, it is necessary to add some balancing points to the above treatment of the resurgence of US power. First of all, it is important to distinguish US imperialism from globalization even though the discourse of neo-liberal globalization, as we saw in chapter 4, acquires a good deal of its sustenance from an American-led view of the world. A central element of contemporary globalization relates to the reality of increased interconnectedness and interdependence, especially visible in the field of cultural exchanges and the use of the internet. Globalization does not by any means need to be treated as somehow synonymous with Americanization and imperialist power. The world is more complex, more unpredictable and more ambivalent. What need to be borne in mind and given continuing analysis as part of a post-colonial geopolitics are:

a) the resurgence of US power as a central element of global times, and the significance within this resurgence of the directing role of the US state rooted in an assertive nationalism;
b) the combination of global interconnectedness with the persistence of North–South divisions, the asymmetry of power relations and the reproduction of subordinating modes of representation.

From a Latin American perspective that validates sovereignty and autonomy, the contemporary influence and impact of US policies are seen in a decidedly negative light. In the context of sovereignty, US–Cuban relations and the drug certification process both illustrate the deployment of a disciplinary power, linked into subordinating modes of representation, which generates sustained opposition within Latin America. The increase of US military assistance to Colombia (e.g. through Plan Colombia, whereby the US has been sending Colombia's armed forces aid valued at $1.5 million per day in 2001) and perceived US complicity in the attempted coup against the democratically elected government of Hugo Chávez in Venezuela in April 2002, also do little to counter the negative memories of past US interventions in Latin America. Moreover, as a measure of growing US militarization in Latin America, in 2000, for the first time since before President Kennedy's 'Alliance for Progress' (discussed in chapter 3 above), total security assistance to Latin America exceeded total economic assistance (Isacson 2001). In addition, post 9/11, some guerrilla organizations such as the FARC (Revolutionary Armed Forces of Colombia) have been reclassified as 'terrorist' groups.

But it is important not to portray Latin American societies as passive subjects in the face of the 'Colossus of the North'. For example, it is of some significance that South American leaders now meet a minimum of five times a year to discuss common problems in associations such as the Rio Group Presidential Summit, the Mercosur Presidential Summit and the Ibero-American Summit. In the past, hemispheric presidential summits were convened by the US, but in the current era Latin American leaders meet regularly and independently to discuss a wide range of issues. In addition, in terms of changing trade relations, there is evidence of growing intra-South American interdependence, so that while exports from the Mercosur countries (Argentina, Brazil, Uruguay and Paraguay) to the NAFTA market fell from 23.9 per cent in 1990 to 16.9 per cent in 1995, intra-Mercosur trade increased from 6.9 to 20.5 per cent (Muñoz 2001: 80–1). Of more relevance for my own perspective, in the field of oppositional movements and calls for new ways of thinking about global politics, Latin America has been a creative region for alternative ideas, as we shall see in chapter 8.

Finally, returning to the comment made by Fernando Henrique Cardoso concerning the proposal that Latin American societies publish reports evaluating the policies and actions of the US – the call for a process of 'writing, and theorizing back', discussed in chapter 5 in relation to *dependencia* – it is worthwhile underlining that aspect of post-colonial thinking that enables us to take the critique back to the heartlands of global power. In the context of the spread of US influence, especially in relation to the notion of exporting democracy, a good place to start would be in the state of Florida during the 2000 presidential election, where defective voting machines, gerry-mandering and chicanery came together to form a salient democratic deficit which was not salvaged by the decision of the US Supreme Court.[12] In other words, the actual prerogative of diffusing one model of democracy to the societies of the South, needs to be challenged by going back to the limitations of the 'model society' and posing questions to that uncritically enframed model, as did members of the Russian Duma, who voted a resolution demanding that American presidential elections, like Haiti's and

Rwanda's, should be held under the auspices of the United Nations (Palast 2003: 80).

Resisting the exported model, challenging the existing state of global affairs, arguing for another kind of world, lead us into a new set of questions – post-colonial questions for global times, part II.

8

'Another World is Possible': On Social Movements, the Zapatistas and the Dynamics of 'Globalization from Below'

'We are diverse – women and men, adults and youth, indigenous peoples, rural and urban, workers and unemployed, homeless, the elderly, students, migrants, professionals, peoples of every creed, colour and sexual orientation. The expression of this diversity is our strength and the basis of our unity. We are a global solidarity movement, united in our determination to fight against the concentration of wealth, the proliferation of poverty and inequalities and the destruction of our earth.'
– Social Movements at the World Social Forum, 2002, Porto Alegre (www.portoalegre2002.org; accessed 11 Feb. 2002)

Learning from the South: Social Movements and Issues of Representation

A recurrent theme of Western writing on the societies of the global South has been to associate their realities with that of societal chaos. Huntington (1998: 321), for example, in discussing contemporary threats to 'civilization', notes that law and order in much of the world – Africa, Latin America, the former Soviet Union, South Asia, the Middle East – appears to be evaporating, while the world as a whole would seem to be heading towards a global Dark Ages. In other texts, the Third World or

South is represented in terms of a perilous frontier land, or as the setting for the emergence of new wild zones in an increasingly dangerous world. Clearly, in some countries of the South, especially perhaps on the African continent, war and social division and the breakdown of central state authority can be taken as one part of the current social and political reality, a reality the causes of which are partly linked to the effects of neo-liberal policies. But what needs to be stressed here is that this kind of essentialized representation, where the societies of the South are used as an immediate example of chaos, breakdown, violence and overall dystopia, draws a veil over a much more complex reality, within which there are positive tendencies from which we can learn.

In the sphere of social-movement analysis, essentialized depictions of the periphery have also surfaced. Sometimes, for example, it has been suggested that the peoples of the Third World are not able to achieve a decisive advance in the handling of the new questions of women's liberation and ecology, since they are confronted with the still potent fetters of traditional custom, including the persistence of patriarchy, and with pressing problems of material need that make it tempting to neglect qualitative long-term concerns such as the reproduction of ecological systems (see Otto Wolf 1986: 37). In other instances, the Third World has been associated with grandiose political programmes that have been seen to result in violence and totalitarianism.[1] The point here would be that it is important *not* to engage in a politics of simple reversal whereby (paraphrasing Stuart Hall) we would replace 'the essential, bad, white (First World) social subject' with the 'essential, good, black (Third World) subject';[2] instead, it is crucial to recognize diversity, openness and the plurality of agency and subjectivity.

Hence, for example, whereas it is important to examine the meanings and historical contextualizations of revolutionary politics in the periphery, or the varied and complex significance of the use of violence in social action, or the particularity of cultural traditions in influencing the content of social struggles, equally, in the context of the global South, it is important to analyse, *inter alia*, the trajectories of democratic politics, the emergence of new ways of 'doing politics' and the transcendence of cynicism through collective action and rebellion, or what Subcomandante Marcos (2000) calls the struggle against the 'globalization of pessimism'.[3] In all cases, there will be heterogeneity and difference which contrast with and allow us to break open the 'othering' essentializations of uncritical Western discourses. One part of contesting these essentializations relates to accepting the need to critically learn from the

South, both in terms of the new ways of political mobilization and the innovative forms of reflection and theorization.

In this chapter I intend to develop a post-colonial perspective on questions of social movements and globalization from below in which the primacy of the geopolitical and the pivotal position of the periphery will be highlighted.[4] After discussing some salient aspects of the theorization of social movements, I shall focus on the issue of the spatialities of contemporary social struggle and the pursuit of democratization. In this context, the significance of the Zapatista rebellion and the World Social Forum as a 'global democratic encounter' will be examined as illustrative examples of the specificity of today's alternative global politics.

New Forms of Subjectivity and the Multiplicity of Resistance

At the beginning of the 1980s, the surfacing of large-scale mobilizations against the nuclear arms race, environmental degradation, human rights violations and the persistence of patriarchal relations of power generated a discussion of what became known as the 'new social movements'. This discussion crossed the North–South divide and raised a series of questions concerning identity, difference, democracy and political subjectivity.[5] In an insightful contribution to understanding the new forms of identity emerging within the social movements of this period, the German sociologist Tilman Evers (1985) suggested that in the process of creating new patterns of socio-cultural practice, the individuals as well as the group were developing the fragments of a new subjectivity which was not constituted by a definite end-point but rather was to be seen in the context of fragments of emancipatory subjectivity. Evers saw these movements as the bearers of fragments of a new subjectivity in so far as they had succeeded in overcoming some aspects of alienation and in constructing some initial elements of an autonomous identity. The emphasis that Evers gave to fragmentation and to the interweaving of individual and collective subjectivities reflected the work of feminist theorists who had put the issue of multiple subject positions firmly onto the analytical agenda.

Moving beyond the notion of a unified social subject, it was argued that in each individual there are multiple subject positions corresponding both to the different social relations in which the individual participates and to the discourses that constitute these relations. For example, each social subject is inscribed in a range of social relations connected to gender, race, nationality, production, the environment, locality and so on. All these relations are the basis of subject positions, and every social subject is thus the site of many subject positions and cannot be realistically reduced to one position that is necessarily determining. Moreover, each subject position is itself the site of a plurality of discursive constructions which give different meanings to positionalities around gender, race, nationality and so on. From this line of argument it follows that the subjectivity of a given social agent can never be finally fixed; rather, as was intimated in chapter 6, it is more appropriately envisaged as provisionally and precariously constituted at the intersection of various contrasting discourses (Mouffe 1988).

In this sense, for example, the development of a democratic struggle can be seen as not only the connection of the different mobilizations of more than one movement, for example, a linking up of indigenous and environmental struggles, but also as the pursuit of the democratization of the multiple subject positions of each agent in the struggle for democracy. For example, within the ecological movement there would be a continuous struggle against racist/ethnocentric and sexist/patriarchal positions and not solely against positions causing environmental degradation. Žižek (1990: 250) expresses this point rather well, noting that in so far as the participant in the struggle for democracy discovers through experience that there is no real democracy without the emancipation of women, or in so far as the participant in the ecological struggle finds out through experience that there is no real reconciliation with nature without abandoning the aggressive-masculine attitude towards nature, and in so far as the participant in the peace movement discovers that there is no real peace without radical democratization etc., it is possible to say that a unified or converging subject-position is being constructed; i.e. to be a democrat means equally to be a feminist, an ecologist and a peace activist. Nevertheless, such a potential unity or at least convergence is not the only possible outcome, but rather one possible result of a 'symbolic condensation' around a notion of democratization that is radically contingent. It is one possible construction among several, and it would be quite possible, for instance, to imagine an ecological position that saw the only solution in a strong authoritarian state resuming control over the exploitation of natural resources.

In this context, it is possible to think of movements in terms of the social construction of collective identity. Thus, social subjects produce an interactive and shared definition of the goals of their action and the terrain on which it is to take place. For Melucci (1992) such a definition is an 'active relational process' and collective identity is constructed and negotiated through an activation of the social relationships connecting the members of a group or movement. Here, cognitive frames, dense interactions and emotional and affective exchanges are expressive of the continuing process of the formation of collective identities. This process is found both within a movement and also in the more recent example of a 'movement of movements', where the World Social Forum, originally initiated in Porto Alegre, provides a new space of democratic encounter that spreads out globally, a theme to which I shall return below.

Process, as has been already indicated, is crucial in both the general area of subjectivity and in the formation of collective identity. This basic but sometimes neglected point is also relevant in the analysis of the relations between discourses and social identity. The realization of particular versions of meaning in forms of social organization relies on the discursive constitution of subject positions from which individuals actively interpret the world and by which they are themselves reciprocally affected. Clearly the social and political affinities of any discourse will not be realized without the active agency of individuals who put into practice those affinities and allegiances, and in this way it is possible to suggest that social subjects are both the site and agents of a discursive struggle for identity – both the recipients and protagonists of meaning and action. And in that process of reception and projection, discourses are themselves altered and reconfigured. At the same time, the construction of individuals as subjects is an ongoing process and is always open to challenge and destabilization. Similarly, with oppositional social movements, the questioning of sedimented meanings – as reflected, for example, in the women's movement's interrogation of patriarchal relations – has required the articulation of different subject positions that Foucault (1980b: 100–1) termed new 'reverse discourses'.[6]

It was also Foucault, in his chapter on the deployment of sexuality, that made a number of observations that remain highly relevant to any analysis of the imbrications of power, discourse, resistances and movements. For Foucault, as was briefly mentioned in chapter 1, there is no single locus, source or determining origin of all rebellions, but instead one has a plurality of resistances, each one being a special case. Although there may occasionally be 'massive binary divisions' or 'great radical ruptures', more often one encounters 'mobile and transitory points of resistance, producing cleavages in a society that shift about, fracturing unities and effecting regroupings'. Resistances for Foucault are the odd term in relations of power, inscribed in them as an irreducible opposite, and 'the points, knots, or focuses of resistance are spread over time and space at varying densities, at times mobilizing groups or individuals in a definite way' (Foucault 1980b: 96). How then does Foucault specify these points of resistance or kinds of struggle?

In a later treatment of the subject and power, three kinds of social struggle are identified: (a) struggles against ethnic, social and religious domination, (b) struggles against exploitation that separates individuals from what they produce, and (c) struggles against submission and sub-jection (Foucault 1986: 211–12). Foucault regarded the third type of struggle as becoming increasingly significant, being associated with a number of new or original features.

First, there is a questioning of the place of the individual in society, whereby on the one hand there is the assertion of the right to be different while on the other there is strong opposition to anything which separates the individual or splits up community life. Second, the privileges of official knowledge, its secrecy and disinformation, are strongly chal-lenged, calling into question the established relation of knowledge to power. Third, there is a rejection of state violence and a refusal of a 'scientific or administrative inquisition' which attempts to determine who one is. These innovative features, which are relevant to the 1980s debate on the new social movements, and also to the contemporary discussion of protest and change, revolve around the dynamic intersec-tions among the individual, society and political power. In this context, the crucial political and ethical problem of the day was to try and liberate oneself from both the state and the type of individualization that is linked to the state – what is then needed are new forms of subjectivity. These new forms of subjectivity will vary in content, and such variation will clearly be deeply affected by the historical and societal context. To illustrate this idea and to take the argument forward, I want to take an example from Latin America and specifically from Brazil.

At the beginning of the 1990s, it was noted that in Latin America, in addition to the rooted problems of social polarization, marginalization, underemployment and unemployment, poverty and social deterioration, the continent was facing new problems from emigration and drugs. This seemingly intractable situation led one Brazilian social theorist, namely Weffort (1991), to stress the deleterious effects of social decay, of the dangerous prospects of a generalized anomie, of a loss of a future and of a sense of societal disintegration.[7] Pitted against this dystopian scenario, Weffort (1991: 93) affirmed the force of democracy as the 'force of hope'. And this democracy referred not only to the nature of political power but equally significantly to the extraordinary increase in the organizational capacity of civil society. How then would such an organizational capacity connect to new forms of subjectivity and the proliferation of resistances?

Remaining with the Brazilian case, Evelina Dagnino (1998) provides a revealing opening onto this question. Examining the intersections between culture and politics in relation to social movements, Dagnino suggests that the perception of the need for changes in the country's authoritarian culture was critical to the struggles of, for example, women, gays and blacks, and that this was also part of the struggle for democratization. With the urban popular movements the overlappings of culture and politics became quite clear when these movements realized that their struggles were not only for housing, health, education and so on, but more fundamentally for the right to have rights. Part of the authoritarian, hierarchical ordering of Brazilian society meant that to be poor was not only to endure material deprivation, but to submit to cultural rules that carried a complete lack of recognition of poor people as subjects in their own right. Thus the struggle for the right to have rights established a connection between culture and politics as being crucial to collective action. Furthermore, the fact that for urban popular movements the perception of social needs came to be connected to the belief in the right to have rights generated a new conception of citizenship. This also meant that the kinds of political tactics characteristically accepted by the popular sectors – namely, favouritism and clientelism – came to be less predominant as new forms of subjectivity in relation to citizenship and democratic politics emerged. For Dagnino (1998: 50–1) there are three key points in this context.

First, a new notion of citizenship carries with it a redefinition of the idea of rights, and centrally a conception of the right to have rights. Thus the right to autonomy over one's own body, the right to environmental protection, the right to housing are examples of this formation of new rights, and moreover this redefinition begins to include the right to equality as well as the right to difference. Second, the new sense of citizenship reflects and requires an active definition of rights by the social subjects themselves and is therefore quite different from the traditional notion of excluded social subjects being gradually incorporated into capitalist society according to the dominant power relations. This struggle of the excluded for their own definition of rights and social recognition can be described as a strategy of securing citizenship from below. Third, rather than seeking access to an already given political system, what is at issue here is the right to participate in the actual definition of that system. This has been exemplified in cities governed by the Brazilian Workers' Party, such as in Porto Alegre from 1989 (see Baierle 1998 and Wainwright 2003), in which popular sectors have been engaged in enhancing the democratic control of the state as the participation of citizens in decision-making processes has been consolidated.

I have given some space to the argument developed by Dagnino because it neatly illustrates and carries forward the points schematically outlined by Foucault, showing that in the Brazilian case the struggle against subjection and subordination in society has gone together with a positive affirmation of the right to have rights and the emergence of new forms of subjectivity and visions of a more democratic form of citizenship. Equally, the various social movements discussed by Dagnino show how the 'points of resistance' identified above are linked together through new approaches to 'doing politics' where the material or economic, ethnic, cultural and social are not separate registers but interlinked dimensions in a project for societal renovation.

Having thus sketched out some relevant features of thinking on new forms of subjectivity and the plurality of resistances to the contemporary configuration of power relations, it is now necessary to pose questions concerning the more overtly spatial nature of this contextualization and to do so with reference to the notion of 'globalization from below'.

Inside/Outside and Spaces of Resistance: The Case of the Zapatistas

In chapter 1, a distinction was drawn between politics and the political, which were denoted as two registers that implicate and involve each other. Politics has its own public space – it is the field of exchanges between political parties, of parliamentary and governmental affairs, of elections and representation and in general of the type of activities, practices and procedures that take place in the institutional arena of the political system. The political, in contrast, can be more effectively regarded as a type of relationship that can develop in any area of the social; it is a living movement, what Arditi (1994) calls a kind of 'magma of conflicting wills', mobile and ubiquitous, going beyond but also subverting the institutional moorings of politics. This idea of a magma, of an unpredictable and mobile fluid that disturbs and challenges the sedimented practices of politics and culture, is particularly germane to a more explicitly spatial treatment of politics and the political. In particular, in what can be referred to as social movements that challenge the spatiality of existing political systems, demanding for example a decentralization of power and a territorial extension of democratic politics, there may also be a call for the democratization of power at other spatial levels. Mobility, connectivity, and overlappings come to generate a rethinking of geopolitics and the geopolitical in ways that challenge existing modes of institutional practice. This is clearly the case with the Zapatista uprising in 1994 and its continuing impact.

The armed uprising of over 3,000 indigenous people in the state of Chiapas in south-eastern Mexico on 1 January 1994, and the occupation of seven towns in the highland zone of this state, was timed to coincide with Mexico's entry into the North American Free Trade Agreement (NAFTA) with the United States and Canada. In fact, one of the first communiqués of the *Ejército Zapatista de Liberación Nacional* (EZLN) stated that NAFTA 'is a death certificate for the Indian peoples of Mexico, who are dispensable for the government of Carlos Salinas de Gortari' (qtd in Harvey 1995: 39). The salience of this statement was subsequently reflected in a leaked Chase Manhattan Bank memorandum for early 1995, which asserted that the Mexican government would have to destroy the Zapatistas in order to demonstrate their effective control of the national territory and of security policy.[8] The Zapatistas did not, however, simply express their opposition to NAFTA; they articulated a series of demands and propositions that called into question relations of power across a broad range of spatial levels, the global, the supra-national, the national and the regional and local.

In their first declaration from the Lacandon forest, the Zapatistas issued a comprehensive list of 11 demands relating to work, land, housing, food, health, education, independence, freedom, democracy, justice and peace. Taking their name from the peasant revolutionary Emiliano Zapata (1879–1919), the first declaration included an historical context-ualization of social struggle: 'we are the product of 500 years of struggle'. Specifically, and as one way of legitimizing their uprising, the EZLN cited article 39 of the Mexican Constitution, which states that, 'national sovereignty resides essentially and originally with the people . . . and the people have the inalienable right to alter or modify the form of their government' (qtd in Vázquez Montalbán 1999: 264). At the same time, the Zapatistas treat the national level as one sphere among many,[9] and their claims and proposals do not envisage the national level as separate from the global, supra-national or regional and local. This sense of the overlapping of levels or spheres is partly captured in the following graphic quotation from Subcomandante Marcos talking about Chiapas in a way that echoes much earlier *dependentista* writing, as discussed in chapter 5:

> Chiapas is bled through thousands of veins: through oil ducts and gas ducts, over electric wires, by railroad cars, through bank accounts, by trucks and vans, by ships and planes, over clandestine paths, third-rate roads and mountain passes... Oil, electric energy, cattle, money, coffee, bananas, honey, corn, cocoa, tobacco, sugar, soy, melons, sorghum, mamey, mangos, tamarind, avocados and Chiapan blood flow out through a 1,001 fangs sunk into the neck of southeastern Mexico... Billions of tons of natural resources go through Mexican ports, railway stations, airports and road systems to various destinations: the United States, Canada, Holland, Germany, Italy, Japan – but all with the same destiny: to feed the empire.[10]

This quotation reflects a geopolitical imagination that fuses a variety of spatial levels, the global, the national, the regional and the local. Such a fusion is again evident in an interview with the Zapatista leader published in August 1995, from which three key arguments can be highlighted.

First, it is suggested that current processes of globalization have the potential to break nation-states and to accentuate internal regional differentiations, as reflected in the divergence among the northern, central and south-eastern zones of Mexico. Second, with reference to war, it is noted that political confrontation and the battle for ideas has acquired more significance than military power, echoing the Gramscian contrast between a war of position and a war of movement or manoeuvre (as mentioned in chapter 6) and crucially foregrounding issues of cultural difference and conflict. Third, and following on from the primary importance of the battle for ideas, major emphasis is given to the role of the means of communication, and the need to disseminate alternative visions of indigenous struggles.[11]

It is in this situation that the Zapatistas have been able to find creative ways of exercising control over the representation of their activities in a country where the electronic media are dominated by one mega-network, Televisa. Their effectiveness in disseminating their objectives and ideas has been aided by the solidarity of other agents, particularly sympathetic newspapers such as *La Jornada* in Mexico City, local radio stations, and perhaps above all through transnational media and especially the internet. As Yúdice (1998: 372) points out, the Zapatistas' expert handling of the electronic media proves that there is no necessary contradiction between technological modernization and grassroots mobilization.[12] In fact, their inventive use of the internet, illustrated by an engagement with global issues such as their opposition to NAFTA and neo-liberalism, has encouraged some authors to describe the Zapatista insurgency as a 'post-modern revolutionary movement' (Burbach 2001), or for Castells (1997: 79), as the first informational guerrilla movement. What is important to remember here is that one of the most powerful aspects of the Zapatista rebellion has been their ability to *combine* an effective utilization of the internet with the diffusion of a political perspective that connects to a variety of global issues while being firmly rooted in the local, regional and national specificities of Mexican society.

In the context of the international *élan* associated with the Zapatistas and their overall global visibility, it is important to emphasize their rooted nature. Indeed, as a number of researchers have reminded us, the Zapatista rebellion is anchored in a long regional history of social struggle and opposition which has provided it with a deep cultural and political sustenance (see e.g. Dietz 1995, Harvey 1998, Higgins 2000 and Zermeño 1995). Furthermore, its leadership has expressed a respect for difference and plurality that has displayed a sharp contrast to previous revolutionary movements, and this contrast has taken shape through a process of learning from the indigenous communities of the Chiapas region. This process has included a 'humanization' of political theory, through for instance expressing political ideas in poetic form, and also crucially through a greater realization of the necessity of understanding cultural differences with respect to conceptions of time, nature and life. As a result, new forms of political subjectivity have emerged together with multi-dimensional points of resistance that have not been solely embedded in material and socio-economic questions, although these have been of considerable importance, as noted below. The new forms of subjectivity and resistance clearly need to be connected to the causes of the insurgency, which are anchored in the specificities of Mexican history.[13]

In the regional context, an important point to remember is the heritage of rebellion. The Maya stand out among peoples who strenuously resisted conquest. They were not subjugated in Yucatán and Guatemala until 1703, and they soon rebelled again, organizing a great revolt in 1712. The fact that people from the same region rebelled again in January 1994 was linked to their resistance to the oppression that was undermining their sense of identity and dignity as men and women whose lands were constantly being taken from them. The heritage of rebellion is also associated with the power of myths such as the extant myth of Juan López, an unconquerable man of the Lacandon rain forest and the Highlands of Chiapas, who, according to local belief, came down from heaven to fight the army and who promised to return to help native Mexicans in later battles.

Intimately linked into the history of collective struggles and a consciousness of rebellion, one has contemporary negative trends in relation to land and autonomy. In the Lacandon rain forest, the different ethnic groups – the Tzeltales, Tzotziles, Choles, Zoques, Tojolabales – related to each other and an identity arose among all of them as oppressed ethnic groups facing plantation owners and cattle ranchers. This identity began to emerge in the 1970s and it intensified in the 1980s with the formation of the Rural Association of Collective Interest (ARIC), and culminated at the end of that decade in a coming together of organizations of ethnic groups and workers.

As one example of the negative trends in land control and distribution, the 1971 Presidential decree handed over a very large section of the Lacandon rain forest to an almost extinct ethnic group – the Lacandons. Under the pretext of preserving land for one group, it was intended to deprive other ethnic groups of the land they had inhabited for 20 to 30 years. A company named Forestal Lacandona signed a contract with the new 'owners' and thus acquired the right to extract 35,000 square meters of lumber per annum. Aided by the central government, the company proposed to 'relocate' many of the indigenous people of the area, and while some left the region others began to struggle in defence of their land. Correspondingly, in other parts of the region, an estimated 100,000 people had been forced to emigrate from the central valleys due to the construction of reservoirs, and additionally oil drilling had made large tracts of land unsuitable for cultivation. Further, Chiapas state governors moved to protect large landowners from expropriation by using force in the eviction of peasants who were classified as 'land invaders'. Agrarian disputes became fiercer, and in the early 1980s 400 plantations and latifundia were invaded by peasant farmers. By the 1990s, 27 per cent of unsatisfied demands for land in Mexico as a whole originated in the state of Chiapas (González-Casanova 1995b: 13).

Finally in 1992, the federal government, in agreement with the IMF, and in preparation for NAFTA, decided to amend article 27 of the Mexican Constitution, so that *ejido* and communal land could be privatized, thus providing a legal basis for a new cycle of land concentration (see Harvey 1998: 187). In Chiapas, the implications of this change were seen as particularly threatening not only given the fact that at that time there was still a considerable backlog of unresolved land petitions, but that also land occupations had been met by a violent response organized by the large landowners of the region. Repression had included the imprisonment and assassination of peasant leaders.

These socio-economic and political circumstances provided potent raw material for conflict and rebellion, but equally one ought not to ignore the specificities of the politicization process in the Chiapas region. One important factor originated with the theology of liberation and the location within Chiapas of Catholic priests who, as part of their pastoral contribution, spread through the indigenous communities ideas concerning equality, dignity and democracy. Their educational endeavour was quite beyond anything attempted by a political party or cultural entity. For example, Samuel Ruiz, a bishop of San Cristóbal de las Casas, and with the help of priests, parish curates and deacons, prepared more than 400 pre-deacons and 8,000 catechists in 2,608 communities (González-Casanova 1995b: 6).

Another source of political influence emanated from the student movement. After the 1968 massacre of students in Mexico City, the surviving student leaders took many roads, but some began to reach Chiapas in the mid-1970s. They joined popular organizations, and working together with indigenous groups these students began to change their political outlook, leaving behind more orthodox visions of socialism and the class struggle and replacing them with a greater emphasis on justice, democracy and the struggle against exploitation and oppression. They discovered, as González-Casanova (1995b: 9) puts it, that '"reordering the world" can only come from a struggle for democracy that includes and is based on the autonomies and rights of native peoples and of the poor.'

In November 1983 six individuals (three Indians and three *mestizos*) arrived in Chiapas to found the Zapatista National Liberation Army, and Marcos arrived a year later. For Marcos, through listening to the tales and myths of the local inhabitants and learning from the grounded as well as mythical experiences of the indigenous people of Chiapas, a process of the 'Indianization' of the EZLN began (Higgins 2000: 361). At the same time, through a dialogical encounter the Zapatistas were induced to reconsider their own interpretation of the historical process of the Mexican nation.

The real extent of the 'Indianization' of the EZLN was only made manifest when the indigenous communities decided themselves to go to war. With the creation of the Clandestine Revolutionary Indigenous Committee – General Command (CCRI-CG) in 1993, the Zapatista general command became essentially Indian and the CCRI-CG became the central means of coordination and dissemination for the preparation of the 1994 rebellion. According to Harvey (1998: 224), it can also be noted that in the creation of the EZLN, male-dominated community assemblies were restructured by women's demands for equal participation in the struggle, demands which were reflected in the Zapatistas' Revolutionary Women's Law which stated that all women should have the right to political participation on an equal footing with men.

In the initial weeks of the rebellion many hundreds of lives were lost. Although the Zapatista rebel positions were bombed, the Mexican government refrained from unleashing its full fire-power, not only because of the guerrilla tactics deployed by the Zapatistas but more because of the national and international media attention they had succeeded in acquiring through their dramatic declaration of insurrection (Higgins 2000: 370). The violence used against the indigenous population of Chiapas was not a new phenomenon,[14] and nor was it to recede as subsequent events in the 1990s showed, a notorious example being the killing of 46 people, mainly women and children, in the community of Acteal in late 1997. In general, the uprising led to a militarization of Chiapas, so that from a small presence of 4,000 soldiers in 1987, by 1999 there were an estimated 60,000 to 70,000 soldiers stationed in the region (Vázquez Montalbán 1999: 275).

As mentioned above, and despite the negative effects of increased violence, one of the distinctive features of the Zapatista leadership has been their emphasis on the importance of difference and plurality, allied to democracy, dignity and social justice. In the mid-1990s its 16 popular demands concerning land, housing, work, food, health, education, culture, information, independence, democracy, freedom, justice, peace, security, anti-corruption and environmentalism were articulated in the form of a dialogue and the organization of a national consultation. This national *consulta* took place in August 1995 and elicited a response from approximately 825,000 people, of whom 97.7 per cent approved of the 16 demands. Over 90 per cent were in favour of political reforms, while 56.2 per cent expressed the view that the EZLN should convert itself into an independent political force (reported in *La Jornada*, Mexico City, 29 Aug. 1995: 5). In response, the Zapatista Front of National Liberation (FZLN), an independent political force, was formed at the end of 1995, with the intention of developing a new kind of politics without being involved in the electoral system. In March 1999 the FZLN organized another *consulta* that allowed just under 3 million Mexicans to vote on the unratified San Andrés Accords on indigenous rights and culture, which had been negotiated between the federal government and the EZLN and signed in early 1996.[15] In addition, 30,000 Mexican citizens living in 156 cities around the globe, from Oslo to Patagonia, set up polling stations and participated in the rebel army's second mass sounding of public opinion (*Latinamerica Press*, 12 April 1999: 3). The referendum turned out nearly 3 million votes in favour of the Accords becoming law.

In the case of the Zapatista uprising and its continuing challenge to the existing institutional order in Mexico and beyond, it is possible to suggest that within an internal/national domain a consideration of the geopolitical or a counter-geopolitics, as mentioned in chapter 1, can have two linked meanings.

First, the Chiapas uprising, with its crystallization of deeply rooted social opposition to the existing disposition of power relations with respect, for example, to land distribution, indigenous rights and governmental oppression, can be seen as representing a radical questioning of the territorial functioning of the contemporary Mexican state. Its list of demands, and its prioritization of a radical democratization of society, have been articulated in a context of space and power, connecting the community, municipal/local, state or regional and national spheres in a broader global setting.

Second, the Zapatistas represent a movement that, through its naming, reconnects to one of the founding protagonists of the Mexican Revolution. In a continuing act of radical remembering, the movement reframes the themes of land, justice, dignity and democracy. Through a process of the reactivation of contested meanings, the Zapatistas present themselves as both a recovery and a renewal of a moment of resistance that interweaves the cultural with the geopolitical and the socio-economic with the ethical and philosophical. At the same time, the effects of neo-liberal globalization, and more concretely the impact of NAFTA on national autonomy, provided an example of an externally generated geopolitical moment that impinged on the ensemble of internal spheres, the regional, local and community levels as well as the central state or federal level. Hence, the actual timing of the uprising and the trajectory of Zapatista discourse cannot be understood outside the continuing intersection of inside and outside, of internal and external. This can be further illustrated through reference to the 'First Intercontinental Forum for Humanity and Against Neoliberalism'.

In 1996, in the Fourth Declaration of the Lacandon Rain Forest, the Zapatistas called for a world of many worlds and affirmed that democracy will come when the culture of the nation is refashioned from the perspective of the indigenous peoples. Rooted in such a perspective the Zapatistas reached out to the whole of humanity, and the invitees to their Intercontinental Forum included a broad array of people from collectives, movements, social, citizen, political organizations, neighbourhood associations, cooperatives, non-governmental organizations, groups in solidarity with the struggles of the peoples of the world, bands, tribes, intellectuals, musicians, workers, artists, teachers, peasants, cultural groups, youth movements, alternative media, ecologists, squatters, lesbians, homosexuals, feminists and pacifists. The forum was attended by 3,000 people from 54 countries, and in the subsequent publication one of the key themes concerned the critique of neo-liberalism, which was situated in a local/regional and national/global framework (EZLN 1996).

Since 1996 the Zapatista struggle has focused on the passage of the indigenous rights law, based on the San Andrés Accords. In February 2001, the EZLN launched a march from Chiapas to lobby Congress for passage of the law. The EZLN and National Indigenous Congress (CNI) arrived in Mexico City in March of the same year to a welcome from an estimated 160,000 supporters. However, the Mexican Congress drastically reduced the effectiveness of the law, whereby, for example, the power to legislate for autonomous municipalities was delegated to 31 state legislatures. In addition, the collective use of land and natural resources was taken out of the law, clearing the way for national and transnational corporations to begin extracting forest resources and exploiting mineral rights. The concept of territoriality, which forms a key part of the International Labour Organization's Convention 169, which Mexico had ratified at the end of the 1980s, and which had been written into the San Andrés Accords, was also erased from the law.[16] In the Mexican Senate there appeared to be a fear of a fractionalization of the Mexican republic and the creation of a fourth level of government. Hence, without the specification of the territory within which indigenous rights might be exercised, the right of autonomy was effectively derailed (for a discussion see Higgins 2001). Criticism of these disempowering changes, which effectively stripped the law of all provisions relating to self-determination, was immediate, and within 48 hours the Zapatistas broke off all communication with the government.

Since the passage in 2001 of the Law of Indigenous Rights and Culture the Zapatistas have become more quiescent. On the January 2002 anniversary of their uprising a statement was read out which criticized the government for mutilating the indigenous rights law, while also stating that 'we celebrate the eighth year of our "war against oblivion"' (quoted in *Latinamerica Press*, 28 Jan. 2002: 4). Also, as they celebrated their 8 years of struggle in Oventic, 50 kilometres from San Cristóbal de las Casas, they did so next to a clinic, auditorium, new middle school, library and basketball court which they had won through their struggles for dignity and self-determination. In January 2003, 20,000 masked Zapatistas marched into San Cristóbal, chanting slogans against President Vicente Fox's Plan Puebla to Panama Project and against NAFTA and the projected Free Trade Area of the Americas.[17] The change in political mood was also reflected in the nearly 18-month silence of Subcomandante Marcos, broken in October 2002 in an unexpected manner through his criticisms of the Spanish government's treatment of the Basque independence movement. In Chiapas, pro-Zapatista communities have been evicted from the Montes Azules biosphere reserve in the Lacandon forest, and Vicente Fox has been threatening to build dams that would inundate a significant portion of the EZLN's zone of influence (*Latinamerica Press*, 15 Jan. 2003: 4–5).

Overall, it is possible to discern a downturn in the fortunes of the Zapatistas, and this change has to be linked to their inability to push through a radical indigenous rights law which would have constituted formal recognition for a central aim of their rebellion. This failure underlines the persistence of opposition to the Zapatistas within Mexico's established political party system, with sections of both the PRI and PAN parties sabotaging key sections of the rights law. It also demonstrates the crucial nature of the national level: that a failure here has detrimental effects for state/regional, local/municipal and community levels. This does not detract from the other achievements of the Zapatistas, to which I shall return subsequently, but it does help us to avoid overly romanticized visions of their impact. Also, despite the need to situate the uprising and its trajectory in a global setting, and 9/11 certainly did not aid the Zapatista cause,[18] the unfolding of events shows that the nation-state level remains quite pivotal to any analysis of the varied and complex processes of a counter-geopolitics.

From Chiapas to Porto Alegre: The World Social Forum and Democratic Politics

Moving from the Zapatista uprising to the emergence of the World Social Forum, which has been likened to a movement of movements, enables us to shift the focus to the broader concerns of the organization of an alternative global politics – or a 'globalization from below' – which, as I shall indicate, has close links to the Zapatista rebellion. First, what is the World Social Forum?

The first World Social Forum (WSF) in January 2001 attracted around 5,000 registered participants from 117 countries, plus thousands of Brazilian activists (Teivanen 2002: 624). For five days (25 to 30 January) in Porto Alegre, Brazil, in both the Catholic University of Rio Grande do Sul and the streets, parks and cultural centres of the city a wide range of participants – diverse social movements, labour unions, peasant organizations, indigenous people's organizations, women's movements, NGOs, youth organizations and so on – came together to discuss the negative effects of neo-liberal globalization and the emergence of new kinds of global solidarity. By the following January, at the second WSF, there were over 15,000 registered delegates from social movements, labour unions and NGOs, and over 50,000 participants from 131 countries. The official WSF website was accessed on a daily basis by approximately 550,000 people.[19] By January 2003, the WSF welcomed 20,763 delegates from 5,717 organizations from 156 countries (see *Le Monde Diplomatique*, edición española, no. 88, enero 2003: 2).

The WSF is most appropriately seen as a key part of the wider movement against neo-liberal globalization and its name and time of coming together reflect opposition to the G-8 World Economic Forum meeting in Davos, Switzerland. Initially a collective of Brazilian social movements and organizations, including the Brazilian Workers' Party (the PT), came together with support from the French monthly *Le Monde Diplomatique* and its allied ATTAC (Association for a Tobin Tax to Aid Citizens) and began to explore the possibilities of establishing an alternative forum from 1998 onwards. By 2000, the coordinator of the Brazilian Entrepreneurs Association for Citizenship, a progressive entrepreneurial association, the director of *Le Monde Diplomatique* and the chair of ATTAC met in Brazil to discuss the actual setting up of a new World Forum. It was agreed that the meetings should be held in the South, and specifically in Porto Alegre, that the organization should be called the World Social

Forum, and that it should meet at the same time as the WEF. Subsequently, a number of Brazilian civil society organizations decided to support the initiative, and in early 2000 both the mayor of Porto Alegre and the governor of the state, Olivio Dutra, enthusiastically welcomed the idea. Later that year, the initiative was successfully presented at an alternative meeting of the UN in Geneva by the vice-governor of the state of Rio Grande do Sul (Teivainen 2002: 623–4).

The initiative to establish a World Social Forum was taken in a broad context of world-wide demonstrations against neo-liberal globalization and its driving institutions such as the World Bank, the International Monetary Fund and the World Trade Organization (for a recent analysis of this world-wide movement, see Callinicos 2003). As one example, in 2000 alone, there were mobilizations in Bangkok at the tenth UNCTAD meeting, at the UN in New York and at the headquarters of the World Bank and the IMF in Washington, at an OECD meeting in Bologna, Italy, at a G-8 meeting in Okinawa, Japan, and at an IMF meeting in Prague. In addition, in the same year, rural organizations led by the Peasant Way, and opposed to the impact of neo-liberalism in the countryside, held their Third International Conference in Bangalore, India, where a number of peasant movements participated, including the Brazilian Landless Rural Workers' Movement (MST), the Thai Assembly of the Poor, and the Indian Federation of Peasants.

The year 2000 saw a growing geographical extension and also coordination of these various mobilizations and movements. Furthermore, it needs to be borne in mind that social conflict around the periphery of the capitalist world (for example in Indonesia, Thailand, Korea and India) and especially in Latin America was intensifying. In fact it can be argued that Latin America and specifically the Zapatista rebellion had a creative influence on the overall global movement against neo-liberalism. The Argentinian social scientists Seoane and Taddei (2002: 102) make the point that the First Intercontinental Encounter for Humanity and against Neo-Liberalism that was organized by the EZLN in Chiapas in 1996 (referred to above) acted as a first step in building the international movement against neo-liberal globalization. The Zapatista movement with its opposition to NAFTA appeared as the first major social movement since the fall of the Berlin Wall to represent not only the indigenous and oppressed of Mexican society but also all the world's oppressed peoples. The Zapatista initiative of the 'intergalactic encounters' was extended to Barcelona in 1997 and to Belém in Brazil in 1999, and helped to stimulate the setting up of Global People's Action (GPA) in early 1998.

Also from within Latin America, the Hemispheric Social Alliance, which was formally constituted in 1999, has had an important influence on the setting up of the World Social Forum. The Alliance is a network of citizens' coalitions and labour organizations representing more than 45 million people from across the Americas. It was created to facilitate information exchange and joint strategies and actions towards building an alternative, democratic model of development that would benefit the peoples of the whole region. It is, according to its website (www.asc-hsa.org), an open space for organizations and movements interested in changing the policies of hemispheric integration and promoting social justice in the Americas. Its coordinating committee consists of a range of organizations, including the Mexican Action Network on Free Trade, the Civil Initiative for Central American Integration, the Latin American Congress for Peasant Organizations, the Brazilian Network for the Integration of Peoples, the Canadian organization Common Frontiers and the US Alliance for Responsible Trade. As part of its general principles, it is stated that 'we must find ways to take creative advantage of globalization and not passively submit to it,... and... as citizens of the Americas we refuse to be ruled by the law of supply and demand and claim our role as individuals rather than as simple commodities governed by the laws of the market' (Hemispheric Social Alliance 2001: 6). In this context, the Alliance has set out four guiding principles: democracy and participation; sovereignty and social welfare; the reduction of inequalities; and the promotion of sustainability. These principles are echoed in the World Social Forum's documents, as mentioned below.

The World Social Forum has an open organizational structure with its Organizing Committee consisting of the Central Trade Union Confederation (CUT), the Landless Rural Workers Movement (MST) and six smaller Brazilian civil society organizations. The other key organ of the WSF is the International Council, which was founded in 2001 in São Paulo, and consists of various organizations from different parts of the world. In relation to the Organizing Committee it has an advisory role and no voting power on decisions taken by the Committee. The WSF also has a Charter of Principles which includes the stipulation that to belong to the forum, an organization or group must abjure the use of violence and defend the notion of a globalization of solidarity that respects universal human rights and the environment (quoted in *El País Semanal*, Madrid, 16 March 2003: 59).

The politics of the WSF can be summarized according to five main objectives:

a) the defence of democracy and social justice, not just within the framework of elections, but with a special emphasis on support for the democratization of state and society;

b) the struggle for a new international economic, financial and commercial order, which is more just and equal;

c) the struggle for peace and opposition to militarism and war, whereby conflicts should be resolved through dialogue and negotiation;

d) the defence of the sovereignty and the right to self-determination of the peoples of the world, especially indigenous peoples, with an emphasis on autonomy and social rights;

e) the struggle to preserve biodiversity, connected to the fact that food sovereignty at the local, national and regional level is a basic human right, and that in this respect democratic land reforms and peasant access to land are fundamental necessities.

These objectives need to be placed in the context of resistance to neo-liberalism, war and militarism, which was clearly expressed by the 'call of social movements' at the WSF in 2002 (www.portoalegre2002.org). In this statement, there is a critique of US unilateralism, with its 'walking away from negotiations on global warming, the antiballistic missile treaty, the Convention on Biodiversity, and the UN conference on racism and intolerance.' In addition, other international issues are dealt with, such as the Israel/Palestine question, for which it is stated that 'an urgent task of our movement is to mobilise support for the Palestinian people and their struggle for self-determination as they face brutal occupation by the Israeli state' (ibid.) The 'call of social movements' document concludes by affirming the intention to strengthen the movement through common actions and mobilizations for social justice, for the respect of rights and liberties, for quality of life, equality, dignity and peace.

The desire to engage with international issues and to confront the effects of capitalist globalization have led to suggestions that the World Social Forum represents a new kind of 'International', a 'Rebel International', or a 'People's United Nations' for the twenty-first century. However, the Spanish writer Monereo (2002) points to the differences and conflicts within the movement – for example, not only the differences in objectives and modes of practice that can surface among NGOs, labour unions and grassroots movements, but also the dissonances that have emerged around questions of whether or how far the existing international financial institutions should be reformed or abolished altogether, or whether some regional organizations like UNCTAD should be strengthened (Bello 2002). Furthermore, as Routledge (2003: 344) suggests in respect of his study of Global People's Action, unequal power relations still occur, as reflected in the persistence of patriarchal attitudes and actions in several of the peasant movements that participate in the PGA. These differences and power inequalities ought not to be surprising in a movement of movements that comprises at least five main components: international social movements; international NGOs; labour unions and human-rights networks; centres of research and action; and means of alternative communication such as *Le Monde Diplomatique*. Moreover, there are North–South differences in historical and geopolitical experiences.

One salient difference concerns the levels of violence associated with private-sector or state responses to social-movement activity. One particularly striking example is provided by the violent reaction to the Brazilian Landless Workers' Movement (the MST). Part of the movement's activities to bring about land reform and redistribution in a country with an extremely unequal land tenure system involve putting tens of thousands of people onto the land by means of occupations that last for periods of weeks, months and sometimes years. The landowners' response has often been violent, hiring paramilitary groups or individual hired guns, and there were an estimated 1,167 assassinations of rural workers from 1986 to 1998 (Meszaros 2000: 7). The killings were at their peak in the late 1980s, reaching 161 in 1987, when supporters and opponents of agrarian reform mobilized to exert pressure on the Constituent Assembly (Hammond 1999: 475).

In another example, in Bolivia, during the month of April 2000, when the 'war over water' in Cochabamba reached its initial peak – a conflict sparked by plans to privatize water and cede control of water provision to a transnational consortium (*Aguas de Tunari*) – the government declared a 'state of siege' and during the protests there were 5 deaths and 42 people were seriously injured, and on one day alone as many as 28 people were wounded with one fatality (CEDIB 2000: 2).[20] Also, in Argentina, Mertes (2002) reports that since March 2001 at least 30 protestors have been killed, and similar situations of violent state response to popular mobilizations are widespread in the societies of the South. In contrast, in the North, as a rule, protestors reach home unscathed, and few if any are shot dead by the forces of the state.

It is in this context of the ever-present threat of violent reactions that many social movements in the South have stressed the urgent need for collective organization and grassroots sustainability so that leaders are not isolated from the body of the movement when they can be more easily targeted and eliminated. In equal measure, it has been a trend within some movements to stress collectivity and avoid the cult of the leader – as the Cochabamba trade-union leader, Oscar Olivera, put it, in relation to the struggle against the privatization of water provision in Bolivia: 'the movement needs as many spokespersons as possible and not *caudillos*.'[21]

Another difference relates to the material difficulties involved in sustaining popular movements in which participants may not have the resources to travel to meetings. For example, with respect to the WSF itself, it needs to be remembered that as Mertes (2000) usefully points out, time, money and a constraining sense of distance can present real obstacles to the participation of the rural and urban poor and to activists and workers across Latin America as well as other parts of the South. Materiality is also crucial in issues of land reform, which are still pivotal in societies such as Brazil, and where mobilizations have to be aimed at the national state level, where the essential decision-making power to implement reform is located. It is here too that we have a link with the earlier discussion of the Zapatista rebellion, since there also the national level remains fundamental to the process of radical change. Does this imply that other levels are less important, and how in this context do we think through what has been referred to as 'globalization from below'? (Falk 1993).

Alternative Spatialities of Solidarity

Globalization from above can be thought of as another name for neo-liberal globalization, a process that is founded on privatization, competitiveness, deregulation, standardization and more profoundly the commodification of social life. For Falk (1999) this kind of globalization is predatory and homogenizing, whereas globalization from below is associated with heterogeneity, diversity and bottom-up participatory politics. This distinction can be helpful and avoids the problems associated with a full-blown denunciation of globalization in its entirety, from which the alternative tends towards a somewhat uncritical notion of localization. It can be suggested that while a neo-liberal globalization from above promotes competitiveness, hierarchy, conformity and the primacy of the cash nexus, globalization from below can help expand the ethic of participatory democracy to a variety of spatial levels, not just the global but the supra-national, national, regional, local and community levels. It is not that more power at one level of governance will necessarily disempower people at other levels, but that the empowerment of local and national communities requires the extension of democratic principles at the global and supra-national levels. As Brecher et al. (2000) constructively suggest, globalization from below requires a framework that recognizes this interdependence of spatial spheres or levels. It can also be suggested here that when globalization from above intersects with globalization from below, the point of maximum tension will tend to locate itself at the national level – what I previously referred to as the geopolitical pivot, where the pressures from above and below interact with the most impact.

With reference to the WSF, the above theme can be looked at in terms of the coalescence of a point of arrival and a point of departure. The siting of the WSF in Porto Alegre was a reflection of a range of influences. The governmental presence of the Brazilian Workers' Party at state (regional) and local (municipal) levels was an enabling presence in terms of providing a political space as well as needed financial support, together with progressive civil society associations mentioned above. The setting up of the WSF in Porto Alegre also flowed out of the struggles of social movements in Latin America, the activities of the Hemispheric Social Alliance *and* the global wave of anti-neoliberal mobilizations. These mobilizations, even before Seattle in 1999, were beginning to impact on the decision-making centres of capitalist power, most notably reflected in the derailment of the transnational corporations' MAI (Multilateral Agreement on Investment) initiative, which if implemented would have given greater power to big firms over nation-states. But Porto Alegre and the World Social Forum are not only a point of arrival for these diverse influences and flows of resisting power – they have also become a point of departure for a further broadening of *counter-sites* to globalization from above. This is clearly reflected in the inauguration of an African Social Forum in 2001, which met again in Addis Ababa in 2003 (Robert 2003), and a European Social Forum in 2002, as well as plans to host the 2004 World Social Forum in Mumbai as part of the formation of an Asian Social Forum.

While the World Social Forum's point of arrival and point of departure may be based in the struggles against neo-liberalism, the meetings and activities generated by the WSF as a counter-site of protest and an alternative, counter-hegemonic globalization are motivated by a desire for more democracy, more social justice and more dignity for the peoples of the world. Not only, as the Zapatistas have done, do they place onto the agenda the right to have rights, but also the movements and associations belonging to the WSF are involved in Foucault's three kinds of social struggles, mentioned in the first part of the chapter. But they go further. They are not only against ethnic and religious domination, capitalist exploitation and new forms of subjection and subordination, but they are against new forms of imperialist power, of 'global colonialism', and moreover they have a vision for an alternative kind of global politics based on redistribution and recognition – the drive towards greater equality together with a greater recognition of difference. This requires respect for the autonomy of different movements while seeking out what may be held in common and what might bring movements together in new forms of cooperation. Differences are to be respected, but commonalities discovered. As Santos (2001) suggests, common ground needs to be identified in, for example, an indigenous struggle, a feminist struggle and an ecological struggle, without cancelling out in any of them the autonomy and difference that sustain them. The autonomy and the difference, the commonality and the connectivity, have to be held in creative tension. This can be done through a struggle for greater democratization which includes an articulation of struggles rather than a passive acquiescence of their separation (Laclau & Mouffe 2001).[22]

The idea of a counter-hegemonic globalization, a globalization from below that not only challenges the neo-liberal doctrine of capitalist expansion and a resurgent imperialism, but at the same time offers an alternative vision of how the world could be organized, also can be viewed as offering the possibility of a counter-geopolitics. A transnational project for global justice and participatory democracy which does not prioritize any one spatial level, and does not downgrade the relevance of the national level (Glassman 2001), offers a real alternative to the current hegemony of neo-liberalism. The actual practice of opposition has also been innovative, as the previous director of the World Development Movement, Barry Coates, has pointed out. Face-to-face lobbying, alliance-building, the arrival in politicians' mailboxes of thousands of letters, cards and emails from the public, stories placed with sympathetic journalists, working through trade union and political party structures, and the production of alternative proposals on world trade and investment done through international coordination via the internet, all came together in a successful campaign to block the MAI initiative (see Green & Griffith 2002). A similar campaign is now underway against the attempt to revive the MAI initiative, which is linked to another ongoing campaign against the GATS proposal on the privatization of services (see WDM 2003). But also street protests and demonstrations are a key part of the resistance movement, as was vitally clear on 15 February 2003, when over 8 million people marched on the streets of the world's five continents, protesting against the imminent invasion of Iraq.

The counter-geopolitics that I have invoked above is rooted in an optimism of the will that goes beyond national boundaries, that encompasses activists across borders, and provides a new kind of globalization. It is taken forward by grassroots activists, progressive NGOs, civil society organizations, pressure groups and critical writers and intellectuals like Eduardo Galeano, Walden Bello, Vindana Shiva and Martin Khor. An archipelago of resistances that engenders new spatialities of solidarity and hope for a more emancipatory politics of the future.

9

Conclusions: Beyond the Imperiality of Knowledge

The interweavings of geopolitical power, knowledge and subordinating representations of the other have a long history. For example, the identity and authority of Western modernity took shape on the terrain of colonial and imperial power, and the production of knowledge that characterized the development of Western scientific disciplines went together with the establishment of modern imperialism. In a similar vein, the history of comparative literature, cultural analysis, and anthropology can be seen as affiliated with imperial power, and as contributing to its methods for ensuring Western ascendancy over non-Western peoples. Together with this intertwining of power and knowledge one can locate varying forms of subordinating representation which are equally geopolitical and cultural. The assumption of Western supremacy goes together with a silencing of the non-Western other. There is incorporation, inclusion, coercion but only infrequently an acknowledgement that the ideas of colonized people should be known.

This silencing of the non-Western other is customarily combined with representations that legitimize the power to penetrate and to re-order. The posited superiorities of Western 'progress', 'modernization', 'democracy', 'development' and 'civilization' are deployed to justify a project of enduring invasiveness. The non-Western society is shorn of the legitimate symbols of independent identity and authority, and its representation tends to be frozen around the negative attributes of lack, backwardness, inertia and violence. It becomes a space ready to be penetrated, worked over, restructured and transformed. This is a process that is seen as being beneficial to the re-ordered society, so that resistance, especially in its militant form, is envisaged as being deviant and irrational. So while power and knowledge are combined together, they cannot be adequately grasped if abstracted from the gravity of imperial encounters and the geopolitical history of West/non-West relations.

One of the recurrent themes of this study has been the intersection between intervention and representation. Geopolitical interventions as examined in the nineteenth, twentieth and now twenty-first centuries entail different forms of representation. Nevertheless, it can be argued that they all presuppose a combination of desire, will, capacity and justification. The desire to intervene, to possess, to take hold of another society, even if only temporarily, flows from that deeply rooted sense of superiority and mission. The nineteenth-century notions of 'Manifest Destiny' or 'benevolent assimilation' were predicated on a belief in the ostensible superiority of the Western and more specifically American way of life. It was not just that the United States had a ruling vision of itself that was associated with a destiny that needed to be fulfilled; it was a vision that was also embedded in a hierarchical perspective on peoples, races and cultures, whereby the white/black binary division was seen as a crucial marker of value and significance. As was suggested in chapter 2, in the history of US expansion, race went together with notions of destiny and mission, as exemplified in the US–Mexico War, the colonization of the Philippines and the creation of semi-protectorates such as Cuba at the beginning of the twentieth century. In that era, the desire to intervene was also linked to protecting the Americas from the insecurity of political disorder, and the tenets of civilization were closely associated with the stipulated need to preserve socio-economic order and stability throughout the American hemisphere.

The desire to intervene can also be traced through the histories of modernization theory and neo-liberalism. In the aftermath of the Second World War, modernization ideas were formulated with a view to diffuse Western capital, technology, and social and political values to societies that were judged to be traditional and in need of modern transformation. However, as I suggested in chapter 3, the desire to project modernization was also tied to a fear of the perceived vulnerability of Third World societies to the 'contagion of communism'. Therefore, intervention had a double motive, and while the perceived threat from communism came to a close in 1989, the desire to modernize and re-order the other has continued into the post-Cold War period.

Whereas, as was argued in chapter 4, the neo-liberal doctrine of development can be distinguished from modernization theory in its greater prioritization of the private sector and of the commodification of social and economic life, it shares with modernization a privileging of a certain view of the Western experience in matters not only economic, but also governmental, social and more indirectly psychological. Its desire to intervene was initially anchored in a perception of an economic malaise, of a debt crisis in the Third World that called for the cure of structural adjustment, deregulation and privatization. The 'money doctors' of the international financial institutions have written a series of prescriptions that amount to much more than a case of economic intervention, and their will and capacity, as well as underlying desire, have been systematically extended to cover a broad social, economic and political terrain, stretching from structural adjustment through good governance to social capital.

Desire, as I am using it here, denotes a feeling of unsatisfied longing, a feeling that satisfaction would be derived from obtaining or possessing a given object. Intervention is an action that would facilitate such a satisfaction, but such an action requires both the will and the capacity to realize the desire. If desire is longing for a certain possession or attainment, will represents a kind of concentration of desire. The will to intervene is a focusing of the desire and is reflected in individual, collective and governmental action. Moreover, as with desire, the will to intervene, as reflected, for example, in governmental action, can be envisaged as a condensation of multiple determinants – for instance, cultural, economic, military, political. In both desire and will there is multiplicity, but with will its focusing requires a greater degree of discursive order. Governments, or more broadly states, as well as international institutions such as the IMF or the World Bank, provide focal points for the will to intervene; and equally, as we have seen in the case of the United States, such a will has been allied to a multiple capacity to intervene. The state provides the will and also coordinates the capacity to intervene; and the desire, will and capacity all have a history and a geopolitics (see chapters 2, 3 and 7 above).

Desire, will and capacity, to be effective as an ensemble of meaning and practice, need a language of legitimization. The will to intervene as a crystallization of desire can only be deployed with effect when the capacities – military, economic, political – to intervene are in place. Will and capacity together provide a force, but their power is secured as hegemonic power through the deployment of a discourse of justification. A will that focuses desire, and allies itself to capacity, seeks a hegemonic role through the power of inducing consent while retaining the ability to coerce.

In the context of the imperiality of US geopolitics, the will to power has utilized a connected array of ideas and concepts to ground its projection.

The Roosevelt Corollary of 1904 invoked an international police power, thus underlining the ability when necessary to intervene coercively, but it also incorporated notions of 'civilization' and 'order' that the societies of the Latin South were encouraged to embrace. The 'Good Neighbor Policy' of the 1930s and the Alliance for Progress of 1961 were also concerned with promoting order, but in a context of partnership, cooperation and progress. These were codifications of a will to power that varied according to the geopolitical conjuncture, including the responses to American power from the societies of Latin America. Developing Latin American nationalism in the 1920s and 1930s provided a key backdrop to Washington's framing of a good neighbour policy, and the Cuban Revolution provided a crucial context for Kennedy's announcement of the Alliance for Progress, as well as for the unsuccessful US-sponsored invasion of Cuba a month later at the Bay of Pigs.

The plurality of responses to the 'colossus of the North' reminds us of what can be seen as absent from the suggested combination of desire, will, capacity and legitimization. These are notions that have been used to emphasize a certain projection of power, but such a projection can be interpreted as implicitly denoting an array of passive recipients. From the vantage point of the periphery, and especially the Latin South, there would only seem to be an active, ubiquitous 'outside' which as an analytical perspective can occlude the complexities and heterogeneities of the 'inside'. In this context of the interface between a mutating external power and the dynamic specificities of internal social, economic and political processes it is possible to highlight a number of different tendencies.

In my own analysis, resistances to the geopolitical power of the United States have been linked to nationalism, dependency writing, the Cuban Revolution, the Zapatista uprising and the World Social Forum. I have suggested that these forms of resistance have in part been responses to the constraining impact of external power, but also have sometimes, as with Latin American nationalism and the Cuban Revolution, influenced the future trajectory of US strategy. The challenge to US power is only one response – accommodation and indifference being others – but I have tended to profile this particular response because of its active agency and contrasting driving force. At the same time, with respect to the interface of inside and outside, the main objective has been to focus on the changing configuration of imperial power, while maintaining an analytical space for the active and resistant subjects of that power.[1] In addition, although not analysed in the text, it ought not to be forgotten that the resistant subjects of imperial power are not only located in the periphery, since opposition to US and more generally Western foreign policies also resides historically *within* the West itself (Zinn 1980).

In the context of intervention and representation, it is possible to suggest that there are two contrasting forms of knowledge. There is that knowledge of other peoples, other times and of other spaces, that is the outcome of respect, recognition and careful analysis, whereas there is another kind of knowledge that is intrinsic to a strategy of expansion, belligerence and war. There is, in the spirit of Said (2003), a profound contrast between the will to understand difference for the purpose of co-existence and the mutually sustaining enlargement of horizons and the will to exercise control and dominion over the other. This sharp differentiation is particularly apt for the current, post 9/11 period. For example, as was observed in chapter 7, the resurgence of US imperial power has been closely associated with the rising influence of the protagonists of the Project for a New American Century. Here control and external dominion go together with a belief in the right of the United States to 're-draw the map' of the Middle East. To spread 'democracy' and Western 'civilization' through invasion and re-ordering is allied to a notion of missionary power. Not dissimilar visions were present under the Reagan administration, and the modernization theorists of the 1950s and 1960s not only affirmed the centrality of the West but not infrequently recommended intervention to secure external dominion over regions of the ostensibly recalcitrant Third World.

Imperial knowledge has a continuity. At the end of chapter 2, it was suggested that the imperiality of power and knowledge can be seen as having three intertwined elements. First, there is an acceptance and affirmation of the need for expansion and penetration of other societies and cultures deemed to be less advanced. Secondly, this underwriting of an enduring invasiveness goes together with a belief in the right to impose Western values and modes of organization on the non-West. And thirdly, such an imposition is justified in relation to the posited inferiority of the other, and to a lack of recognition of the independent rights of other cultures and societies, including quite centrally, as was seen in chapter 7, their political-territorial sovereignty. These elements are to be found across a wide spectrum and, post-9/11, their exemplification has received renewed vigour. What then of that other more careful form of knowledge that seeks to understand the value of other peoples and cultures?

For this alternative view, knowledge is connected to emancipation and solidarity rather than tied to regulation and control (Santos 2001). It is reflected across a wide range of social science writing that focuses on the global South, and is also present in movements such as the World Social Forum. It is also to be found in those post-colonial critiques that question the privileging of Western modernity. Furthermore, as I have tried to show in the previous chapters, one of the most enabling and constructive features of a post-colonial sensibility is that it not only challenges the imperiality of knowledge and the Euro-Americanist frames that limit our global understanding, but additionally and crucially it raises the issue of the agents of knowledge – who is speaking, from where, for whom and for what purpose? The vitality of what I have termed 'counter-representations' – the richness of critical thought from the peripheries of the world – has been frequently made invisible, just as previously the depth and dynamism of feminist thinking was ignored by an androcentric gaze. Learning from critical thought produced in the South, as particularly examined in chapters 5, 6 and 8, can help us to go beyond the limits of a Euro-Americanist vision, and more fully appreciate the globality of knowledge. Dependency thinking, Latin American perspectives on the post-modern and recent critiques, in the action and word, of neo-liberal globalization provide different and creative pointers to other ways of thinking beyond imperial knowledge.

However, as there is no singularity of response to imperial power, equally there is no singularity of voices from the peripheries of the world; instead, Third World selves are both 'manifold and multiple' (Escobar 1995: 215). The groundedness of multiplicity can help us avoid the pitfalls of essentialization – romanticizing the resistant other, while bypassing the societal violence, polarization and alienation.[2] Multiplicity can also be disruptively juxtaposed to singular notions of a Third World or a global South, which can present us with an interesting challenge.

For example, there is an analytical tension in discussions of the North–South divide or of First World/Third World differentiations, where, while a continuing idea of a binary split is kept in place by the durability of asymmetrical power relations, at the same time, the differences, fractures and heterogeneities within both sides of the divide can tend to destabilize the meaning of the divide itself. There is a creative tension that flows from keeping a binary divide in place while simultaneously believing in the importance of going beyond its limits. As we saw in chapter 1, the North–South categorization can be and needs to be questioned, without that questioning necessarily preventing the critical use of such a distinction in new circumstances.[3] The different interpretive status of these categories is also reflected in the contrast between post-modern and post-colonial thinking, discussed in chapter 6.

For many writers of a post-modern persuasion, categorizations based on centre–periphery or North–South distinctions are seen as obsolete in a globalizing world of proliferating and unpredictable diversity. Complex differences cannot be straitjacketed into old-fashioned binary splits, especially in a post-Cold War era; overall, plurality is paramount. It is precisely here that writers of a post-colonial affinity offer a counterpoint, with their re-assertion of the relevance of such dichotomous distinctions, especially as related to a stress on the coloniality of power. An overlapping of analytical perspectives, as reflected for example in the Latin American discussion of the post-modern, might then lead us into accepting the possibility of the co-existence of complex differences and positionalities, clearly seen, for example, in the discussion of social subjectivities, with the more direct, binary difference that is embedded in the coloniality or more generally imperiality of power. Rather than an either/or situation, *both* perspectives, including the inevitable tensions between them, might be worked with, so that our critical reservoir of thought might be deepened. The imperial divide does not foreclose the existence of other differences of identity and political preference. These other differences co-exist with, attenuate, complicate, add other layers to the divide without necessarily dissolving it.

Moreover, if we are concerned to extend the reach and effectiveness of critical thinking, then clearly it is not advisable to refer only to the contemporary era. But then how do we set about the task of connecting memory to conceptual preference – how, for example, do we select what is remembered in terms of ideas and perspectives as well as events ? One of the aims of chapter 5 which dealt with dependency ideas was to argue that a critical memory on North–South relations would entail examining the potential contemporary relevance of these ideas, as well as their original contextualization in the geopolitical world of their time. Being critical of dependency thinking does not have to lead to its entombment. The selection of dependency perspectives for analysis was connected to the objective of sourcing different critical currents, so that through a process of comparison and synthesis a more linked-up interpretation of North–South relations could be formulated, albeit focused on one type of such relations, namely, US–Latin American encounters. Consequently, dependency provided one kind of analytical terrain for the subsequent treatment of the post-modern, differential elements of Marxist thought and post-coloniality. Here, a suggestion of their nature as an archipelago of critical thinking provided a basis for looking for the possible forms of articulation and connection among them. And again in chapters 3 and 4, the histories of modernization and neo-liberal thinking were not only contextualized geopolitically, but also they were situated as part of a critical memory of ideas – ideas that still have their effects on the direction of North–South relations.

Thus the geopolitics of memory is not only relevant to events and historical circumstances but to the ideas and theories that have been drawn around them. Colonization of the imagination or the reproduction of ruling memories need to be contested through the deployment of critical memory that re-invigorates previous oppositional imaginations as well as re-examining past events that official narratives sometimes erase or belittle. Representations of history, as the Haitian anthropologist Trouillot (1995) observes, need to be associated with some conscious relation to the knowledge they transmit, but critically it is the renewal of the practices of power and domination that should concern us most, since although these practices are rooted in the past, it is only in the present that the struggle against them can be genuinely authentic.

Making the link between past and present so as to profile the need for contemporary struggles against injustice and inequality is a key aspect of the repertoire of a critical or counter memory. It acts as a counterpoint to that other memory that seeks to re-assert older colonial visions of Western supremacy.[4] A struggle over the meaning and significance of past ideas and events is expressive of a clash not of civilizations but of geopolitical world-views. Struggles against present-day injustices and oppression raise another theme involving power and knowledge – namely, the issue of an ethics of responsibility.

From a starting point that states that an ethic of responsibility, allied with a search for a more emancipatory and democratic politics, must include a responsibility to the other, then it follows that there will be a call to resist domination, exploitation, oppression and all other similar conditions that seek to subordinate the other. This would provide a substantive criterion according to which one must mobilize opposition to, for example, ethnic and nationalist chauvinism, fascism, racism, xenophobia, dictatorship and the imperiality of power. However, while such mobilizations might helps us go beyond the ossifying influence of political indifference, nevertheless, in prioritizing struggles against existing conditions, the question of struggles for some wider objective is left unattended. It is here that a notion of radical democracy can be introduced; not as a goal that is restricted to the national level, but as a political horizon that can connect overlapping global, supra-national, national, regional and local spheres in the sense that justice and equality and representative and direct democracy can be fought for and extended through a series of struggles, as has been occurring in Latin American societies such as Bolivia, Brazil, Mexico and Venezuela. In these examples and elsewhere, democratic struggles for greater equality, just-ice, and socio-economic and political rights crucially unfold at the na-tional level, which remains pivotal to political change, even though this level is radically affected by those other levels or domains 'above it' and 'below it'.[5]

To what extent is it possible to interpret concepts such as equality, justice, rights, democracy and solidarity in a universal context, wherein equivalence would outweigh difference as an organizing principle? From an anti-essentialist standpoint, it can be suggested that establishing a universal foundation for deciding what ought to be included within a concept of, for example, justice or democracy will always produce a new site of contestation. This does not mean that there is no foundation, but that wherever there is one, there will also be a foundering or political contest. At the same time, one can posit the existence of an 'ungrounded ground', where foundations or norms or universal prescriptions can be put into question as a permanent feature of the process of democratiza-tion. To challenge universalist treatments is not to engage in their negation but rather to release them from limiting frames of conceptual-ization. This is not to be politically indifferent, but nor is it to insist on a settled ensemble of meanings and practices; rather, it is to mobilize signifiers such as justice and democracy for alternative interpretations and interventions. How do these ideas reconnect to the theme of geopol-itical encounters from a post-colonial perspective?

The general phenomenon of the subversion of territorial sovereignty, which is a vital component of imperial politics, raises a key question for the way notions of justice and democracy are framed in an international context. The state that subverts, such as the United States, is both a democratic and an imperial state. The concept of an imperial democracy draws attention to the intersections of inside and outside, disrupting self-contained and uncritical visions of democracy and justice, while also generating a series of other questions about the inner limits of the democratic system with regard to race and social exclusion. However, this kind of contextualization does not have to carry with it a non-critical position on the state whose sovereignty has been subverted. In such examples, it is important to render problematic either/or options, just as with knowledge and representation, the questioning of dominant Western readings of the non-West does not have to entail an uncritical reading of the non-West itself. Many writers and social scientists from the non-West or global South combine a critical stance against the reproduction of imperialist knowledge with an interrogation of the repressive nature of many Third World regimes. Equally they produce analyses of the socio-political problems of their own societies as well as making vital contributions to the development of critical knowledge – contributions, it needs to be added, that are heterogeneous and multi-layered.

In my exploration of geopolitics, power and knowledge, I have argued that in the broad context of North–South relations, and in particular with respect to US–Latin American encounters, a series of representations have been deployed as a legitimizing mechanism for a continuing power over societies judged to be less civilized, less ordered, less modern and less democratic. Such power is rooted in an imperial subjectivity which entails a subordinating mode of knowledge. It is also a power that seeks to transfer responsibility to the other, to delegate ownership of the process of change, but within a frame that is already established: according to guidelines that are set down as part of a hegemonic project, but which may well be resisted and modified or displaced. In contrast, an ethic of responsibility to the other, to the self and to their inter-relationship, requires going beyond an imperial consciousness.

Such a journey might include developing a perspective which:

a) questions those conceptualizations of, *inter alia*, civilization, order, modernity, development and sovereignty that occlude the geopolitics of power over non-Western societies and represent the West as a self-contained entity, comprehensible outside a context of colonial and imperial relations;

b) profiles the dynamic and relational nature of power through which colonizer and colonized, globalizer and globalized are continually, although differentially, affected in a two-way process;

c) emphasizes the importance of learning critically from those other post-colonial writers and theorists who have represented and continue to propel forward intellectual life beyond and within the West;

d) signals, as suggested in chapter 1, the underlying centrality of the Third World periphery in the formation of the modern world;

e) rethinks the place of an ethico-political sensibility which contests the imperiality of power and knowledge, while taking sustenance from feminist ethics in discussions of (post-)development and knowledge (see e.g. Lazreg 2002 and Saunders 2002).[6]

Such a perspective needs to be genuinely global in the sense that it would not be limited by a Euro-Americanist frame, and it would take care to regard other cultures with respect and recognition, seeking out whatever webs of reciprocity may be located and sustained. There will always be the need for a negotiation of respect and recognition with a critical spirit that helps us avoid essentializations of either a positive or negative hue. A meaningful ethics of intersubjectivity would include the right to be critical and different on both sides of any 'cultural border'. And the future *can* be post-imperial; another world *is* possible.

Notes

Chapter 1: For a Post-colonial Geopolitics

1   Sutcliffe draws our attention to the fact that these data have been criticized for not using criteria such as the weight of children and weight to height ratios, which might well give a more accurate picture than calorie intake (2001: 48).

2   For a detailed analysis of these debates see Slater 1994, and also chapter 6 below.

3   I have discussed Euro-Americanism in more detail elsewhere, and the following section is a revised version of my discussion in Slater 1992, 1999b and 2003a. For earlier critiques of Eurocentrism, see *inter alia* Amin 1989, Blaut 1993 and Shohat & Stam 1994

4   More recently than Husserl, Habermas (1987: 360) has taken a similar position, suggesting that the 'normative content of modernity' has to be analysed in the context of the universal tendency of the European Enlightenment.

5   The term 'America' has a number of different meanings and is not synonymous with the United States of America, hence the inverted commas. I shall take up this theme in chapter 2.

6   See the Cuban Liberty and Democratic Solidarity (LIBERTAD) Act of 1996, Public Law 104-114, 12 March 1996, 110 Stat. 785-824, Washington DC, United States General Printing Office, pp. 786–8.

7   In the case of Britain, one can think of the example of Ireland where subordinating modes of representation accompanied imperial rule on a territory adjacent to the colonizing heartland (see Perry Curtis 1997), so the distinctiveness of the United States in this context is not absolute and of course connections between imperial rule and subordinating forms of representation have been ubiquitous. The specificity of the United States lies in the fact that it was in the process of its own domestic formation that these encounters took place.

8   One of the particularly useful points made by Spurr concerns the way Western powers have imposed their own modes of classification, conceptualization

and thematic prioritization on non-Western societies so that other forms of classifying knowledge appear to be redundant or irrelevant. I shall return to this in chapters 3 and 4.

9   For two contrasting views of this particular problematic see Grovogui 1996 and Zacher 2001. Grovogui, in his focus on the African context, connects questions of international law and state practice to the realities of Western colonialism and the nature of ethnocentric representation that have influenced international legal discourse.

10   In particular the work of Ó'Tuathail and Dalby has been central to the development of a 'critical geopolitics' in which analyses of space, power and politics have been formulated in an imaginative theoretical-empirical manner – see for example Ó'Tuathail & Dalby 1998. On social movements and geopolitics, see, for example, Routledge 2003, and for a general and up-to-date overview of the field of political geography and critical geopolitics, see Agnew, Mitchell & Toal 2003.

11   And of course there will always be debate and difference concerning which authors and which analytical currents are examined and how. In the case of this study I have concentrated my attention on authors from the South who have been and continue to be critical of Western power and its conceits. This is an approach that has its own pitfalls, not least since it does not attempt to be a comprehensive reading, but rather illustrative of trends and issues.

12   In addition, and in another context, Carver (1998) has shown in his book on *The Postmodern Marx* that some of Marx's views on democracy were less economistic than is sometimes realized, giving more emphasis to agents over structures.

Chapter 2: Emerging Empire and the Civilizing Powers of Intervention

1   Already in the Declaration of Independence of 1776, there was a passage that accused the King of Great Britain of aiding and abetting 'domestic insurrections', and endeavouring to 'bring on the inhabitants of our frontiers, the merciless Indian Savages, whose known rule of warfare, is an undistinguished destruction of all ages, sexes and conditions'. See Maidment & Dawson 1994: 157.

2   For the citations from Jefferson's message to the Senate and House of Representatives, see Richardson 1896: 352–4.

3   For an interesting discussion of the more informal 'agents of manifest destiny' – not only influential businessmen and politicians, but the actual leaders of expeditions – the 'filibusters' (freebooters or pirates, the term having both a Spanish (*filibustero*) and Dutch (*vrijbuiter*) origin) who landed

on foreign soil in the Caribbean and Central America to conquer a country with only a few armed men – see Brown 1980.

4   There were of course voices raised in opposition to the dominant sentiments, especially from within the labour movement, Church groups and the anti-slavery movement. One leading representative of the anti-slavery movement, Frederick Douglass, commented in 1848 that United States policy in Mexico was immoral and dishonest, and that there should be an instant recall of US forces – he remarked that 'Mexico seems a doomed victim to Anglo-Saxon cupidity and love of dominion' (quoted in Graebner 1968: 235).

5   My use of the word 'natural' is prompted by the fact that in 1859 US Secretary of State W. Seward told a Spanish minister that Cuba must ultimately come to us 'by means of constant gravitation', also commenting that Cuba had been formed by the Mississippi in washing American sand into the Gulf – see Weinberg 1963: 234.

6   However, at the same time, as Healy (1970: 218–19) has shown, there were non-racialist orientations within the anti-imperialist movement, as reflected, for example, in the labour movement.

7   In 1918 the third of President Wilson's 14 points on peace and democracy advocated 'the removal, so far as possible, of all economic barriers and the establishment of an equality of trade conditions among all the nations consenting to the peace and associating themselves for its maintenance' – see Maidment & Dawson 1994: 259.

8   The quotations are taken from the Government Printing Office, Washington (1943), *Addresses and Messages of Franklin D Roosevelt*, Senate Document no. 188 of the 77th Congress, 2nd Session, and reprinted by His Majesty's Stationery Office, London, 1943, pp. 5–6.

9   These formative elements are customarily associated with the external projection of power, but as was seen earlier on in the chapter, they could also be applied to the internal territorial projection of power over the indigenous peoples of the United States and the Hispanic and African American others in the nineteenth century.

Chapter 3: Modernizing the Other and the Three Worlds of Development

1   For a critical survey of the different currents of modernization theory, see Banuri 1990. There is an extensive literature on modernization, modernity and post-modernity; for an overview see Alexander 1995a and 1995b.

2   It is worthwhile indicating here that one of the differences between modernization theory and earlier visions of Western progress concerned the issue of transformation; the idea of progress, as Bury (1955) noted, was based on an interpretation of history that regarded men as slowly advancing in a definite

direction – the idea carries a sense of incrementalism, whereas in modernization theory there is an ethos of abrupt change and transformation.

3   For some interesting observations on the links inside the United States between the use of the term 'primitive' and treatments of the non-Western or third world, see Torgovnick 1991: 244–8.

4   For an early critique of the geography of modernization, see Slater 1974. For a recent assessment of the literature on development, within which special attention is given to the importance of spatiality, see Crush 1995, and more recently, Peet 1999.

5   This quotation is to be found in Nederveen Pieterse's (1991: 11) article on 'developmentalism'.

6   The Psychological Strategy Board was established in 1951 as part of US psychological strategy in the Cold War, and its objective was to coordinate the policies of the State Department, the Central Intelligence Agency (CIA), the military services and other government agencies. It was judged not to have been successful and was closed down after two years in operation – see Lucas 1996.

7   For a general analysis of containment strategy, see Gaddis 1982, and for an imaginative discussion of its geopolitical impact, see Dalby 1990 and Ó'Tuathail & Agnew 1992.

8   The two Kennan papers are to be found in the Etzold & Gaddis (1978) collection; I have discussed the question of containment in a little more detail in Slater 1999a.

9   George Kennan was a strong advocate of covert action, which he described as including propaganda, economic warfare, subversion against hostile states and support of indigenous anti-communist elements – see Cingranelli 1993: 143.

10 For an evaluation of the role of the CIA in the 1965 events in Indonesia, see Blum 1986: 217–22.

11 This is not to imply that the United States in general and the CIA in particular were not active in other arenas, as was shown for example in Ghana in 1966 with the ousting of Kwame Nkrumah (Blum 1986: 223–5). For a general analysis of the role and place of the CIA in Africa in the post-1946 period see Moore 1996.

12 There is no space here to go into a discussion of the different types of violence, but I would follow Salmi's (1993: 16–23) approach, which broadly defines violence as any avoidable action that constitutes a violation of a human right or which prevents the fulfilment of a basic human need.

13 Similarly, one of Pool's colleagues, Lucien Pye, a founder of counter-insurgency, and in the 1960s Chair of the US SSRC Committee on Comparative Politics, wrote in 1966 of the need to protect a traditional society from the 'calculated attempts by well organized enemies of freedom to use

violence to gain totalitarian control of vulnerable societies' – quoted in O'Brien 1979: 62.

14   In an interesting passage, Huntington (1968: 338–9) compared Madison to Lenin, suggesting that the parallels between *The Federalist* and *What Is To Be Done?* were rather striking since both formulated principles upon which political order could be constructed, Lenin in relation to classes and the party, Madison in relation to factions and representative government.

15   For an intriguing investigation of the 'gang as an American enterprise' see Padilla 1992.

16   The figures on direct US investments in Latin America are to be found in Niess 1990: 206.

17   One of the earliest and most effective critiques of the sociology of modernization and development was made by Frank in the late 1960s – see Frank 1969: 21–94. Subsequently, Taylor (1979) provided a detailed evaluation of modernization ideas and their location in mainstream social theory.

18   The notion of 'upgrading' comes from Talcott Parsons (see Alexander 1995a: 68), and the emphasis on 'rationalization' can be traced back to Weber. The stress on the importance of industrialization and technology can be found in an early paper by Marion Levy (1952).

19   As an example, Jalée (1969: 117) calculated that in 1964–5 the total drain on the Third World by the imperialist countries was equivalent to one and a half times the value of foreign aid.

20   The inclination towards stressing the singularity of Western modernity has been modified by Eisenstadt (2000) in a recent paper on 'multiple modernities'. For an interesting exploration of alternative modernities see, for example, Watts 1996 and the special issue of *Public Culture*, 11(1) (1999).

21   One should also add that modernization theory emerged in an era when Keynesian economics and the 'welfare state', certainly in Western Europe, were still predominant features of the social and economic landscape.

22   The resurgence of the use of the term 'civilization' will be examined in chapter 7, below.

Chapter 4: The Rise of Neo-liberalism and the Expansion of Western Power

1   For a detailed review of the trade liberalization aspects of Bolivia's implementation of neo-liberal policies, see Jenkins 1997.

2   According to a report of the Bolivian Chamber of Deputies in 1987, 'the illegal trafficking of coca leaves and its by-products has become the most dynamic sector in the economy... over the past five years exports of cocaine and its derivatives were 300 per cent higher than all other export earnings put together.' See *Latin American Regional Reports – Andean Group*, RA 87/03, 9 April 1987, p. 7.

3   In the Bolivian case societal acceptance of neo-liberal doctrine was aided by a number of prominent intellectuals; the historian José Luis Roca, for example, argued that the 'democratization of capital' could be partly achieved through the privatization of state enterprises – in *Temas en la Crisis*, issue on *Privatización* (La Paz), no. 39, abril 1991: 43, La Paz. For an alternative perspective, which challenges the connection between democracy and privatization, see Oporto 1991.

4   In December 1990, the Peruvian President, Fujimori, stated that 'we cannot overcome the economic and fiscal crisis without reducing the state machinery and making it more efficient' – surplus bureaucrats would have to go, as would surplus workers (*Andean Group Report*, 31 Jan. 1991). In late 1991, in Argentina, the Menem government announced a far-reaching programme of privatization of state enterprises and liberalization of the economy, and overall it has been estimated that from 1990 to 1995 Latin American countries sold more than US$60 billion in state enterprises, with particularly aggressive privatization thrusts in Argentina, Chile, Colombia, Mexico and Peru – see *Latin America Press* (Lima), 29(22), 12 June 1997: 1. For an insightful analysis of the advent of neo-liberalism in Latin America, see Green 1995.

5   Quite clearly, the studies referred to in this passage only represent one small part of an extensive literature. For two recent and quite comprehensive analyses which go up to the early and mid-1990s, see Chossudovsky 1997 and George & Sabelli 1994.

6   No mention was made of the increase in poverty in Latin America during the 1980s. Data collected by ECLAC showed that from 1980 to 1989 the percentage of the Latin American population living in poverty increased from 41 to 44 per cent – see ECLAC 1991: 66.

7   Laski added here that it was sufficient commentary on the claims of businessmen to hold sway in the economic domain to note that in the United States in 1939 there were 12 million citizens dependent upon relief for their means of life, and that in Britain, the second richest country, one out of every four children was undernourished (Laski 1946).

8   As referred to in Colin Gordon's (1991: 15) introductory essay on 'governmental rationality'.

9   In a quote from Keynes, Hirschman gives us a revealing example of his thesis – Keynes wrote in the 1930s that 'dangerous human proclivities can be canalised into comparatively harmless channels by the existence of opportunity for money-making and private wealth' (Hirschman 1981: 134).

10   The development economist Amartya Sen, in discussing the connections between concepts of the individual and concepts of the economy, once noted that the purely economic man, the ruthless maximizer, in the neo-classical vision, is 'indeed close to being a social moron' (qtd in Williams 1990: 66).

11   The origin of structural adjustment is also associated with the 1981 World Bank report on *Accelerated Development in Sub-Saharan Africa* (the so-called Berg report) – see Ferguson 1995.

12   For one concrete study which focuses on the actual failure of World Bank structural adjustment policies in Africa, see the useful article by Schatz (1996). For a general consideration of Latin American cases see e.g. Demmers et al. 2001 and Green 1995.

13   Conversely, 'unsuccessful decentralization' is defined as a threat to economic and political stability, being disruptive of the delivery of public services – see World Bank 2000a: 107.

14   For an analysis of the influence of neo-classical economic concepts in the social sciences, including the association between human and social capital, see Fine 1999.

15   Putnam leans quite heavily on the sociologist James S. Coleman for his definition of social capital – Coleman defined social capital as being, like other forms of capital, 'productive', making possible the achievement of certain ends that would be unattainable in its absence (quoted in Putnam's discussion of social capital: Putnam 1993: 167–85).

16   A key adviser to Ronald Reagan described the Contras as resistance fighters, comparable to the French Resistance that fought the Nazis – this was the case since Nicaragua was portrayed as a country 'occupied' by a Marxist regime (Kenworthy 1995: 71). In 1989 the US Congress approved an aid package for the Contras of US$67 million (Robinson 1996: 235).

17   Schumpeter (1987: 269) defined the democratic method as 'an institutional arrangement for arriving at political decisions in which individuals acquire the power to decide by means of a competitive struggle for the people's vote'. This definition was taken up by Samuel Huntington, a long-standing adviser to Washington Administrations – see e.g. Huntington 1984. For a detailed critique of US foreign policy and its relation to the framing of ideas on democracy, see Robinson 1996.

18   The example of the Reagan Doctrine underlines the importance of the intertwining of geopolitical and geo-economic considerations and also points to the relevance of viewing changes in the economic strategy of Third World states, as a result of concerted military, political and economic pressure from Washington.

Chapter 5: Societies of Insurgent Theory: The *Dependentistas* Write Back

1   Fanon's remarks on the past under colonial rule are somewhat reminiscent of Walter Benjamin's (1992: 247) famous comment that even the dead will not

be safe from the enemy if he (*sic*) wins, and therefore acquires the power to rewrite history according to his (*sic*) own interests.

2 Nietzsche's (1983: 57–123) argument on the uses and disadvantages of history for life centred around the need to avoid the deadening hand of the past on present and future action and freedom.

3 For a recent and concise review of some of these original ideas, see dos Santos 1998: 54–5.

4 It needs to be noted here that the Economic Commission for Latin America, which was referred to as CEPAL in Spanish (Comisión Económica para América Latina) was later renamed the Economic Commission for Latin America and the Caribbean (ECLAC).

5 These ideas had been worked out in considerable detail in a previous publication dealing with development and underdevelopment – see Furtado 1964.

6 Kay (1989: 88–100) has provided a quite comprehensive overview of these early discussions in Latin America.

7 Also in a Mexican context, González-Casanova (1969) developed a more detailed treatment of internal colonialism, suggesting that it was possible to distinguish three components – monopoly and dependence; relations of production and social control; and culture and living standards.

8 In another passage relevant to this theme, Rodney (1972: 88) writes that European technical superiority did not apply to all aspects of production, but the advantage which the Europeans possessed proved decisive in a few key areas – for example, West Africans had developed metal casting to a fine artistic perfection in many parts of Nigeria, but when it came to the meeting with Europe, beautiful bronzes were less relevant than the crudest cannon.

9 For a useful examination of US military power in Central America during the 1960s in which a similar approach to Ianni's is outlined, see Saxe-Fernández 1969.

10 It was also unfortunate that the Cardoso and Faletto text originally published in Spanish in the late 1960s was not published in English until 1979, by which time the perception and interpretation of dependency ideas had been quite strongly moulded by Frankian notions which were rather less analytically rigorous than those found in the Cardoso and Faletto study.

11 There is no opportunity here to pursue the Warren thesis, and in any case this has been well-debated in the past. In the mid-1980s I discussed Warren's econocentrism and Western ethnocentrism in Slater 1987.

12 For example, it was noted that in the early 1980s the inflow of capital fell well below the level of debt service – see Lipietz 1987: 222 n. 62.

13 In fact this has been a subject much neglected by most currents of development theory. An interesting exception can be found in the work of the Peruvian psychotherapist Rodríguez Rabanal (1989), who in the 1980s

studied the psychological effects of continuing poverty on a number of families in an inner-city area of Lima.

14   Not all the military interventions of these years had the same ideological orientation, and for instance the 1968 military take-over in Peru was characterized by a much more populist and also anti-imperialist strategy than was the case with the military coups in the Southern Cone. For one analysis of the Peruvian case, see, for example, Slater 1989, esp. ch. 5.

15   These and related theoretical and political questions are taken up again in chapters 6 and 8.

16   By the end of the 1990s Brazil continued to suffer from one of the most unequal distributions of income in the world – for example, see UNDP 1998: 29.

17   In a recent paper Kay repeats his earlier point, noting that many misinterpretations of dependency perspectives still persist at the end of the 1990s – see Kay & Gwynne 2000.

Chapter 6: Exploring Other Zones of Difference: From the Post-modern to the Post-colonial

1   Previously, I have looked at this theme in more detail – see e.g. Slater 1994 and 2003b.

2   For a critical note on this aspect of Vattimo's perspective, see Reigadas 1988: 139.

3   In the context of feminist theory, Soper (1991: 99–101) has raised a series of related questions.

4   In the original, Hopenhayn uses the phrase 'cerco neoliberal', which I have rather blandly translated as 'frame'. The Spanish expression is in fact sharper, denoting a sense of being fenced in by a certain mode of thinking.

5   For a detailed analysis of the history of economism in Marxist thought see e.g. Laclau & Mouffe 2001, ch. 1.

6   Ahmad (1995: 16), for instance, remarks that post-coloniality is also a matter of class, while Dirlik (1994) re-asserts the centrality of capital for post-colonial analysis, and Miyoshi (1993: 728) rather polemically asserts that the academic preoccupation with post-coloniality looks like 'another alibi to conceal the actuality of global politics'. For an excellent review of some of these issues see Hall 1996. I have also discussed the politics of the post-colonial in Slater 1998.

7   And here one can be reminded of a comment made by Sartre, where it is noted that the Marxist method 'has already formed its concepts; it is already certain of their truth; it will assign to them the role of constitutive schemata' (Sartre 1963: 37). This quotation is referred to in Lazreg's (1994) study of the history and politics of women in Algeria.

8 And as was suggested earlier on in this chapter, from the critical currents of the Latin American discussion of the post-modern, we can learn how to challenge that modern and post-modern Western conceit of always situating itself on the cutting edge of theory.

9 For particularly imaginative analyses of the links between feminist theory and post-colonial concerns, see, for example, the earlier work of Mohanty (1988) and Spivak (1988 and 1990). For a recent discussion that connects to political geography, see Sharp (2003).

Chapter 7: Post-colonial Questions for Global Times

1 Mosquera is not alone in voicing such arguments, with their stress on questions of power and hegemony in the field of art and cultural production – see e.g. Fisher 1995, Kapur 1997 and Yúdice 1996.

2 The G-77, which takes its name from the alliance of the 77 countries of the South that first took part in the meeting of the UN Conference on Trade and Development in 1964, continues to form an umbrella body for the Southern countries to discuss social and economic questions.

3 General Assembly Resolution 1514 (XV), 14 Dec. 1960, qtd in Philpott 1995: 366.

4 For an explanation of 'quasi-states' and their quasi-sovereignties see Jackson 1993 – and for a critique see Doty 1996.

5 The likening of Cuban communism to a cancerous growth in the western hemisphere appeared in the statements of Republican members of the sub-committee dealing with the western hemisphere in the mid-1980s – see Kenworthy 1995: 130.

6 As one example, in May 2003 the Cuban government made a formal complaint to the Secretary General of the International Union of Telecommunications concerning the use by the United States of a military plane to transmit via its TV Martí station programmes that interfered with Cuba's own channels, violating international regulations on telecommunications – reported in *El País*, 24 May 2003: 6.

7 This charge was further reinforced in 1998 when the United States, without the knowledge of the Mexican authorities, set up a sting operation, code named Operation Casablanca, to trap Mexican banking officials who were suspected of being money launderers. Four days after the arrests in the United States the Mexican president Ernesto Zedillo emphatically rejected the right of any country to carry out undercover operations in Mexico, stating that 'it is inadmissable because it is a violation of sovereignty' – quoted in *Latin America Press* 30, no. 21, 4 June 1998: 1.

8 And it can be added here that the desire of the US to take care of other peoples' sovereignty is a continuing theme; Paul Wolfowitz, for example,

writes of the 'oppressed' Cuban people that 'we have an obligation to deliver the support we promise them', something, according to Wolfowitz, Kennedy failed to do at the Bay of Pigs – see Wolfowitz 2000: 40.

9   On democracy and its spread throughout the world, the US Agency for International Development commented that US support for democracy 'became a central pillar of US foreign policy during the mid-1990s'. This was part of a historic shift from the containment of communism to support- ing democratic expansion as a major goal of American foreign policy (www.usaid.gov/democracy/center/history.html, accessed 3 Oct. 2001).

10   In a recent report released by the US Attorney General's Office, it is admitted that the rights of hundreds of mainly Middle Eastern and South-east Asian origin immigrants have been violated since September 11, being held under suspicion and without trial – see *El País*, Madrid, 4 June 2003: 8.

11   See www.newamericancentury.org/statementofprinciples.htm, accessed on 7 Oct. 2002.

12   While George W. Bush Jr won Florida by fewer than 500 votes, 204,600 Black Floridians were legally barred from voting, and additionally many other of all races who tried to vote could not because their names appeared on felon lists (see Palast 2003: 11–81 and Wilson Gilmore 2002: 273). For a critical review of the decision taken by the US Supreme Court see Dersho- witz 2001.

Chapter 8: 'Another World is Possible': On Social Movements, the Zapatistas and the Dynamics of 'Globalization from Below'

1   This particular view is taken from Melucci's (1989: 189) book on *Nomads of the Present*. It ought to be noted that in his later work and in particular in his 1996 book on contemporary social movements, this earlier view has been replaced by a much more nuanced position – see Melucci 1996.

2   For this reference and other pertinent observations, see Julien & Mercer 1988.

3   In a related remark, Marcos (2001b: 87), in a letter to the Portuguese novelist José Saramago, comments that here in Mexico, we say that both cynicism and rebelliousness fall on to the earth, but only rebelliousness gives us a tomorrow (my translation). Marcos's perspective provides a good antidote to the cynicism of reason mentioned in chapter 1.

4   I should note here that I am using the term 'periphery' as a shorthand for global South, and in particular the Latin South. There are of course other usages of the term 'periphery', but I use it here very much in a North–South context – see chapter 1.

5   For an original and general theoretical treatment of the newness of the new social movements see Laclau 1985, and for an early discussion of new social movements in Latin America see Mainwaring & Viola 1984.

6   Foucault (1980b: 100–1) usefully suggested that 'we must make allowance for the complex and unstable process whereby discourse can be both instrument and an effect of power, but also a hindrance, a stumbling block, a point of resistance and a starting point for an opposing strategy.'

7   Similarly, but in a Mexican context, Zermeño (1989) argued that marginality in the barrio ought to be seen in terms of the universe of egoism, the war of all against all, of a delinquent conformism, of resentment, withdrawal and of an individualistic imaginary nourished by the media.

8   A copy of the Chase memorandum was obtained by the London-based *Independent*, and in a connected passage, the memo stated that while the insurrection in Chiapas 'does not pose a fundamental threat to Mexican political stability, it is perceived to be so by many in the investment community'. See L. Doyle, 'Did US Bank Send in Battalions against Mexican Rebel Army?', *The Independent*, London, 5 March 1995: 14.

9   De Angelis (2000: 20–1) interestingly suggests that Zapatista 'nationalism' has a number of expressions, including the fundamental belief in the nation as a whole that indigenous communities ought to be a part of, and the belief that a 'national interest' needs to be related to humanity and not to capital.

10   The quotation is taken from a work distributed in 1992 entitled "The First Wind", translated by Bardacke & López (1995: 32–3).

11   This interview with Subcomandante Marcos was published in *La Jornada*, Mexico City, 27 Aug. 1995: 10–11.

12   It might be added here that already by the mid-1990s it was noted that Subcomandante Marcos was a prolific writer, and once written his letters and essays were often on the internet by the next morning – see Rich 1997: 74–5.

13   Most of the following paragraphs on the causes of the Zapatista rebellion are based on the Mexican sociologist Pablo González-Casanova's 1995b paper. Also see Harvey 1995 and 1998.

14   According to González-Casanova (1995b: 24), from 1974 to 1987, 982 leaders were assassinated in one part of the indigenous region of Chiapas; 1,084 campesinos were arrested without legal cause; 379 were seriously wounded; 505 were kidnapped or tortured; 334 disappeared; 38 women were raped; thousands were expelled from their homes and land; and 89 villages with dwellings were burned and crops were destroyed.

15   In September of 1996, the EZLN abandoned talks with the Zedillo government because there had been no attempt to implement the San Andrés Accords, and in particular no progress was made to alter the Mexican Constitution in order to include the right to autonomy of the Indian peoples

and autochthonous cultures of Mexico. See Vázquez Montalbán 1999: 281–2.

16   According to the ILO Convention 169 and the San Andrés Accords of 1996, the territory of indigenous peoples is defined as the material base of their reproduction as a people and expresses the indissoluble unity of 'man–land–nature'. See Vázquez Montalbán 1999: 283.

17   While in El Salvador to discuss the PPP (Plan Puebla Panama), President Vicente Fox had made clear his view of the Plan, noting that its importance was 'a thousand times more than *Zapatismo* or an indigenous community in Chiapas' (qtd in Higgins 2001: 901).

18   Of note here is the fact that in January 2002, in an interview at the World Social Forum in Porto Alegre, the state governor for Chiapas, Pablo Salazar Mendiguchia, when asked about the Zapatistas and armed conflict, replied that 'since September 11, . . . armed movements in whatever country of the world are going to have a very difficult future' (my translation). See Icaria Editorial 2002: 27.

19   These figures are taken from Icaria Editorial 2002: 130–1, the source being the State Government of Rio Grande do Sul. The geographical distribution of delegates showed that Brazil was the largest contributor (just over 70 per cent of the delegates), followed by Italy, Argentina, France, Uruguay and the US.

20   More recently, protests at austerity measures introduced by the Sánchez de Lozada government were met with increased violence, and over 50 deaths were reported across Bolivia in February of 2003. See *Latinamerica Press* 26 Feb. 2003: 2, and also Chávez 2003.

21   Interview with Oscar Olivera, 21 July 2000, Cochabamba, Bolivia.

22   Fisher & Ponniah 2003 take up some of the ideas found in Laclau & Mouffe 2001, and give a useful definition of the re-invention of democracy, relevant to the principles of the World Social Forum.

Chapter 9: Conclusions: Beyond the Imperiality of Knowledge

1   For a series of relevant reflections and arguments on the question of imperial encounters and the overlapping of influences as between Latin America and the United States, see Joseph, Legrand & Salvatore 1998.

2   As one example, Briceño-León & Zubillaga (2002: 19) note that by 1998 violence was the leading cause of death among people in the 15 to 44 age group in Latin America and the Caribbean, and it was the fourth most important cause of death for the population as a whole.

3   Coronil (1995: xli), for instance, notes that the issue is not that categories such as First World and Third World should not be used, since in certain

contexts they are indispensable and efficacious, but rather that their use should include an awareness of their limits and effects.

4 For example, in the immediate aftermath of the events of September 11, Italian Prime Minister Berlusconi underlined what for him was the superiority of Western values, and Francis Fukuyama, in his 'end of history' perspective, declared that 'the West had Won'. Both statements, in different but related ways, represent the re-activation of older Western views of imperial privilege which are becoming more widespread. See *The Independent*, London, 11 Oct. 2001: 5, and *The Guardian*, London, 9 Oct. 2001: 24.

5 And it needs to be stated that analytical illustrations taken from Latin America may well not be fully representative of the global South as a whole, since, for example, in this case of the prioritization of the national level, in parts of Africa where there has been continuing political conflict and wars within states, for example in the Democratic Republic of the Congo, Liberia, Sierra Leone and Somalia, such a 'national level' may not always be so solidly present. For a discussion of some of these issues concerning governance and the 'new wars' see Duffield 2001.

6 And it is important to indicate here that in cultural geography, post-colonial issues have been receiving increased attention, often linked to feminist theory – see, for example, King 2003, McEwan 2003, Robinson 2003 and Sharp 2003. For an earlier discussion of the relevance of post-colonial thinking for geographical research, see Gregory 1994, and more recently Sidaway 2000.

References

Abrahamsen, R. (2000). *Disciplining Democracy*, Zed Books, London and New York.

Actualidad Latinoamericana (2000). Año VI, no. 66 (Abril), Instituto Internacional del Desarrollo, Madrid.

Adas, M. (1990). *Machines as the Measure of Men*, Cornell University Press, Ithaca and London.

Agnew, J. (1983). An Excess of 'National Exceptionalism': Towards a New Political Geography of American Foreign Policy, *Political Geography Quarterly* 2(2): 151–66.

Agnew, J. (1999). Mapping Political Power Beyond State Boundaries: Territory, Identity and Movement in World Politics, *Millennium* 28(3): 499–521.

Agnew, J., Mitchell, K. and Toal, G., eds (2003). *A Companion to Political Geography*, Blackwell, Oxford.

Aguilar, A. (1968). *Pan-Americanism: From Monroe to the Present – A View from the Other Side*, Monthly Review Press, New York and London.

Ahmad, A. (1995). The Politics of Literary Postcoloniality, *Race and Class* 36(3): 1–20.

Alexander, J. C. (1995a). Modern, Anti, Post and Neo*, *New Left Review* 210: 63–101.

Alexander, J. C. (1995b). *Fin de Siècle Social Theory*, Verso, London and New York.

Altvater, E. (1998). Theoretical Deliberations on Time and Space in Post-Socialist Transformation, *Regional Studies* 32(7): 591–605.

Amin, S. (1973). *Neo-Colonialism in West Africa*, Penguin Books, Harmondsworth.

Amin, S. (1976). *Unequal Development – An Essay on the Social Formations of Peripheral Capitalism*, Harvester Press, Sussex.

Anderson, P. (1998). *The Origins of Postmodernity*, Verso, London and New York.

Annan, K. (1999). Two Concepts of Sovereignty, *The Economist*, 18 Sept., London, pp. 81–2.

Apter, D. E. (1987). *Rethinking Development*, Sage Publications, Beverly Hills, CA, and London.

Arditi, B. (1994). Tracing the Political, *Angelaki* 1(3): 15–28.

Aricó, J. (1992). Rethink Everything (Maybe It's Always Been This Way), *NACLA Report on the Americas* 25(5): 21–3.

Arndt, H. W. (1989). *Economic Development – The History of an Idea*, University of Chicago Press, Chicago and London.

Arrighi, G. (1991). World Income Inequalities and the Future of Socialism, *New Left Review* 189: 39–65.

Ashley, R. K. (1987). The Geopolitics of Geopolitical Space: Toward a Critical Social Theory of International Politics, *Alternatives* 12: 403–34.

Baber, Z. (2001). Modernization Theory and the Cold War, *Journal of Contemporary Asia* 31(1): 71–85.

Bacevich, A. J. (1999). Policing Utopia, *The National Interest* 56 (Summer): 5–13.

Badie, B. (2000). *The Imported State – The Westernization of the Political Order*, Stanford University Press, Stanford.

Bagú, S. (1970). *Tiempo, Realidad Social y Conocimiento*, Siglo XXI Editores, Mexico and Madrid.

Bagú, S. (1992). *Economía de la Sociedad Colonial*, Editorial Grijalbo, México DF (first published in 1949).

Baierle, S. G. (1998). The Explosion of Experience: The Emergence of a New Ethical-political Principle in Popular Movements in Porto Alegre, Brazil. In Alvarez, S., Dagnino, E. and Escobar, A., eds, *Cultures of Politics, Politics of Cultures*, Westview Press, Boulder and Oxford, pp. 118–38.

Bales, K. (1999). *Disposable People: New Slavery in the Global Economy*, University of California Press, Berkeley and London.

Bambirra, V. (1974). *El Capitalismo Dependiente Latinoamericano*, Siglo XXI Editores, México DF.

Banaji, J. (1973). Backward Capitalism, Primitive Accumulation and Modes of Production, *Journal of Contemporary Asia* 4: 393–413.

Banuri, T. (1990). Development and the Politics of Knowledge: A Critical Perspective on Theories of Development. In Marglin, F. A. and Marglin, S. A., eds, *Dominating Knowledge*, Clarendon Press, Oxford, pp. 29–72.

Bardacke, F. and López, L., eds (1995). *Shadows of Tender Fury: The Letters and Communiqués of Subcomandante Marcos and the Zapatista Army of National Liberation*, Monthly Review Press, New York.

Bartra, R. (1991). Mexican Oficio: The Miseries and Splendors of Culture, *Third Text* 14 (Spring): 7–15.

Baudrillard, J. (1975). *The Mirror of Production*, Telos Press, St Louis, Missouri.

Baudrillard, J. (1990). *Cool Memories*, Verso, London and New York.

Baudrillard, J. (1993a). *The Transparency of Evil*, Verso, London and New York.

Baudrillard, J. (1993b). *Symbolic Exchange and Death*, Verso, London and New York.

Baudrillard, J. (1994). *The Illusion of the End*, Polity Press, Cambridge.

Baudrillard, J. (1996). *Cool Memories II*, Polity Press, Cambridge.

Baudrillard, J. (1998). *Paroxysm*, Verso, London and New York.

Bauman, Z. (1999). *In Search of Politics*, Polity Press, Cambridge.

Bauman, Z. (2002). *Society under Siege*, Polity Press, Cambridge.

Beck, U. (2000). *What is Globalization?* Polity Press, Cambridge.

Bell, D. (1991). The 'Hegelian Secret': Civil Society and American Exceptionalism. In Shafer, B. E., ed., *Is America Different?* Clarendon Press, Oxford, pp. 46–70.

Bell, M. (1994). Images, Myths and Alternative Geographies of the Third World. In Gregory, D., Martin, R. and Smith, G., eds, *Human Geography*, Macmillan, London, pp. 174–99.

Bello, W. (2002). Pacific Panopticon, *New Left Review* 16: 68–85.

Benjamin, J. R. (1990). *The United States and the Origins of the Cuban Revolution*, Princeton University Press, Princeton, NJ.

Benjamin, W. (1992). *Illuminations*, Fontana Press, London (this edn first published in 1973).

Berry, A. (1997). The Income Distribution Threat in Latin America, *Latin American Research Review* 32(2): 3–40.

Bhabha, H. K. (1994). *The Location of Culture*, Routledge, London and New York.

Bhabha, H. K. (1995). The Commitment to Theory. In Carter, E., Donald, J. and Squires, J., eds, *Cultural Remix*, Lawrence & Wishart, London, pp. 3–27.

Bhabha, H. K. (1996). Unpacking My Library Again. In Chambers, I. and Curti, L., eds, *The Post-colonial Question*, Routledge, London and New York, pp. 199–211.

Bienefeld, M. (1994). The New World Order: Echoes of a New Imperialism, *Third World Quarterly* 15(1): 31–48.

Biersteker, T. J. (1990). Reducing the Role of the State in the Economy: A Conceptual Exploration of IMF and World Bank Prescriptions, *International Studies Quarterly* 34: 477–92.

Bilgin, P. and Morton, A. D. (2002). Historicising Representations of 'Failed States': Beyond the Cold-War Annnexation of the Social Sciences? *Third World Quarterly* 23(1): 55–80.

Binder, L. (1986). The Natural History of Development Theory, *Comparative Studies in Society and History* 28(1): 3–33.

Black, G. (1988). *The Good Neighbor*, Pantheon Books, New York.

Black, J. K. (1991). *Development in Theory & Practice*, Westview Press, Boulder and Oxford.

Blaney, D. L. (1996). Reconceptualizing Autonomy: The Difference Dependency Theory Makes, *Review of International Political Economy* 3(3) (Autumn): 459–97.

Blaney, D. L. and Inayatullah, N. (2002). Neo-Modernization?: IR and the Inner Life of Modernization Theory, *European Journal of International Relations* 8(1): 103–37.

Blaut, J. M. (1993). *The Colonizer's Model of the World*, Guildford Press, London and New York.

Blomström, M. and Hettne, B. (1984). *Development Theory in Transition – The Dependency Debate*, Zed Books, London.

Blum, W. (1986). *The CIA: A Forgotten History*, Zed Books, London and New Jersey.

Bonura Jr, C. J. (1998). The Occulted Geopolitics of Nation and Culture: Situating Political Culture within the Construction of Geopolitical Ontologies. In Ó'Tuathail, G. and Dalby, S., eds, *Rethinking Geopolitics*, Routledge, London and New York, pp. 86–105.

Bourdieu, P. and Wacquant, L. (1999). On the Cunning of Imperialist Reason, *Theory, Culture & Society* 16(1): 41–58.

["The Brandt Report"]: Report of the Independent Commission on International Development Issues. (1980). *North–South: A Programme for Survival*, Pan Books, London.

Brecher, J., Costello, T. and Smith, B. (2000). *Globalization from Below – The Power of Solidarity*, South End Press, Cambridge, MA.

Briceño-León, R. and Zubillaga, V. (2002). Violence and Globalization in Latin America, *Current Sociology* 50(1): 19–37.

Broad, R. (1990). *Unequal Alliance – The World Bank, the International Monetary Fund and the Philippines*, University of California Press, Berkeley and London.

Brockway, T. P. (1957). *Basic Documents in United States Foreign Policy*, Van Nostrand Company, Princeton, NJ, and London.

Brown, C. H. (1980). *Agents of Manifest Destiny: The Lives and Times of the Filibusters*, University of North Carolina Press, Chapel Hill.

Brzezinski, Z. (1956). The Politics of Underdevelopment, *World Politics* 9(1): 55–75.

Brzezinski, Z. (1997). *The Grand Chessboard*, Basic Books, New York.

Burbach, R. (2001). *Globalization and Postmodern Politics*, Pluto Press, London.

Bury, J. B. (1955). *The Idea of Progress*, Dover Publications, New York (originally published in 1932).

Butler, J. (2000). Restaging the Universal: Hegemony and the Limits of Formalism. In Butler, J., Laclau, E. and Žižek, S. eds, *Contingency, Hegemony, Universality*, Verso, London, pp. 11–43.

Buzzanco, R. (1999). *Vietnam and the Transformation of American Life*, Blackwell, Oxford.

Caldeira, T. P. R. (1996). Fortified Enclaves: The New Urban Segregation, *Public Culture* 8: 303–28.

Calder, B. J. (1978). Caudillos and *Gavilleros* versus the United States Marines: Guerrilla Insurgency during the Dominican Intervention, 1916–1924, *Hispanic Historical Review* 58(4): 649–75.

Calderón, F. (1987). América Latina: Identidad y Tiempos Mixtos o Cómo Tratar de Pensar la Modernidad sin Dejar de Ser Indios, *David y Goliath: Revista del Consejo Latinoamericano de Ciencias Sociales* 52 (Sept.): 4–9.

Callinicos, A. (2003). *An Anti-Capitalist Manifesto*, Polity Press, Cambridge.

Campbell, D. (1992). *Writing Security: United States Foreign Policy and the Politics of Identity*, University of Minnesota Press, Minneapolis.

Campbell, D. (1996). Political Prosaics, Transversal Politics, and the Anarchical World. In Shapiro, M. J. and Alker, H. R., eds, *Challenging Boundaries*, University of Minnesota Press, Minneapolis, pp. 7–31.

Campbell, D. (1999). Contradictions of a Lone Superpower. In Slater D. and Taylor P. J., eds, *The American Century: Consensus and Coercion in the Projection of American Power*, Blackwell, Oxford, pp. 222–40.

Cardoso, F. H. (1977). The Consumption of Dependency Theory in the United States, *Latin American Research Review* 12(3).

Cardoso, F. H. and Faletto, E. (1979). *Dependency and Development in Latin America*, University of California Press, Berkeley, Los Angeles and London. (Expanded and emended version of the 1969 Spanish text.)

Carey, J. C. (1964). *Peru and the United States 1900–1962*, University of Notre Dame Press, Indiana.

Carver, T. (1998). *The Postmodern Marx*, Manchester University Press, Manchester.

Castells, M. (1997). *The Power of Identity: The Information Age – II*, Blackwell, Oxford.

Castells, M. and Laserna, R. (1989). La Nueva Dependencia: cambio tecnológico y reestructuración socioeconómica en Latinoamérica, *David y Goliath: Revista de CLACSO (Consejo Latinoamericano de Ciencias Sociales)* año XVIII, no.55 (Julio): 2–16.

Castoriadis, C. (1990). Does the Idea of Revolution Still Make Sense? An Interview with Cornelius Castoriadis, *Thesis Eleven* 26: 123–38.

Castoriadis, C. (1991). *Philosophy, Politics, Autonomy* – Essays in Political Philosophy, Oxford University Press, New York and Oxford.

Castoriadis, C. (1998). *El Ascenso de la Insignificancia*, Ediciones Cátedra, Madrid.

Ceasar, J. W. (1997). *Reconstructing America*, Yale University Press, New Haven and London.

CEDIB (Centro de Documentación e Información, Bolivia). (2000). *30 Días de Noticias*, April, Cochabamba, Bolivia.

Césaire, A. (2000). *Discourse on Colonialism*, Monthly Review Press, New York.

Chaliand, G. (1978). *Revolution in the Third World*, Penguin Books, Harmondsworth.

Chávez, W. (2003). La Rebelión Boliviana , *Le Monde Diplomatique*, edición española, año VII, no. 91 (mayo): 8–9.

Chomsky, N. (1993). *What Uncle Sam Really Wants*, Odonian Press, Berkeley.

Chomsky, N. (2000). *Rogue States*, Pluto Press, London.

Chomsky, N. (2003). 'Recovering Rights': A Crooked Path. In Gibney, M. J., ed., *Globalizing Rights*, Oxford University Press, Oxford, pp. 45–80.

Chossudovsky, M. (1997). *The Globalisation of Poverty – Impacts of IMF and World Bank Reforms*, Zed Books/Third World Network, London and Penang.

Chossudovsky, M. (1998). Global Poverty in the Late 20th Century, *Journal of International Affairs* 52(1): 292–311.

Cingranelli, D. L. (1993). *Ethics, American Foreign Policy and the Third World*, St.Martin's Press, New York.

Clapham, C. (1999). Sovereignty and the Third World State, *Political Studies* 57: 522–37.

Conaghan, C. M., Malloy, J. M. and Abugattas, L. A. (1990). Business and the 'Boys': The Politics of Neoliberalism in the Central Andes, *Latin American Research Review* 25(2): 3–30.

Connolly, W. E. (2001). Cross-State Citizen Networks: A Response to Dallmayr, *Millennium* 30(2): 349–55.

Cornelius, W. A. (1986). The 1984 Nicaraguan Elections Revisited, *LASA Forum* 16(4): 22–8.

Coronil, F. (1995). Introduction to F. Ortiz, *Cuban Counterpoint – Tobacco and Sugar*, Duke University Press, Durham and London, pp. ix–lvi.

Coronil, F. (1996). Beyond Occidentalism: Toward Nonimperial Geohistorical Categories, *Cultural Anthropology* 11(1): 51–87.

Coronil, F. (1997). *The Magical State*, University of Chicago Press, Chicago.

Cottam, M. L. (1994). *Images and Intervention – US Policies in Latin America*, University of Pittsburgh Press, Pittsburgh and London.

Cottam, M. L. and Marenin, O. (1999). International Cooperation in the War on Drugs: Mexico and the United States, *Policing and Society* 9: 209–40.

Cox, M. (2001). Whatever Happened to American Decline? International Relations and the New United States Hegemony, *New Political Economy* 6(3): 311–40.

Crush, J. (1995). Introduction: Imagining Development. In Crush, J., ed., *The Power of Development*, Routledge, London, pp. 1–23.

Cullather, N. (1999). *Secret History – The CIA's Classified Account of its Operations in Guatemala, 1952–1954*, Stanford University Press, Stanford, CA.

Cumings, B. (1999). The American Century and the Third World, *Diplomatic History* 23(2): 355–70.

Dagnino, E. (1998). Culture, Citizenship and Democracy: Changing Discourses and Practices of the Latin American Left. In Alvarez, S., Dagnino, E. and

Escobar, A., eds, *Cultures of Politics, Politics of Cultures*, Westview, Boulder and Oxford, pp. 33–63.

Dalby, P. (2003). Reconfiguring 'the International': Knowledge Machines, Boundaries and Exclusions, *Alternatives* 28(1) (Jan.–Feb.): 141– 66.

Dalby, S. (1990). *Creating the Second Cold War*, Pinter Publishers, London.

Dalby, S. (1999). Globalisation or Global Apartheid? Boundaries and Knowledge in Postmodern Times. In Newman, D., ed., *Boundaries, Territory and Postmodernity*, Frank Cass Publishers, London and Portland, pp. 132–50.

David, S. R. (1992/3). Why the Third World Still Matters, *International Security* 17(3) (Winter): 127–59.

Davis, M. (2001). *Late Victorian Holocausts*, Verso, London and New York.

Davis, M. and Sawhney, D. N. (2002). *Sanbhashana*: Los Angeles and the Philosophies of Disaster. In Sawhney, D. N., ed., *Unmasking LA: Third Worlds and the City*, Palgrave, New York, pp. 21–45.

De Angelis, M. (2000). Globalization, New Internationalism and the Zapatistas, *Capital and Class* 70: 9–35.

De Castro, J. (1969). Introduction: Not One Latin America. In Horowitz, I. L., de Castro, J. and Gerassi, J., eds, *Latin American Radicalism*, Vintage Books, New York, pp. 235–48.

De la Campa, R. (1999). *Latin Americanism*, University of Minnesota Press, Minneapolis and London.

De Maillard, J. (2003). Los 'Aliados' en Primera Línea para Proteger el Imperio, *Le Monde Diplomatique*, Edición Española, año VII, no. 87 (enero): 6.

Deleuze, G. and Guattari, F. (1984). *Anti-Oedipus*, Athlone Press, London.

Deleuze, G. and Guattari, F. (1988). *A Thousand Plateaus*, Athlone Press, London.

Demmers, J., Fernández Jilberto, A. E. and Hogenboom, B., eds (2001). *Miraculous Metamorphoses: The Neoliberalization of Latin American Populism*, Zed Books, London.

Der Derian, J. (1992). *Antidiplomacy*, Blackwell, Oxford.

Derrida, J. (1976). *Of Grammatology*, Johns Hopkins University Press, Baltimore.

Derrida, J. (1992). *The Other Heading: Reflections on Today's Europe*, Indiana University Press, Bloomington and Indianapolis.

Dershowitz, A. M. (2001). *Supreme Injustice: How the High Court Hijacked Election 2000*, Oxford University Press, Oxford and New York.

Deutsch, K. W. (1963). Social Mobilization and Political Development, *American Political Science Review* 55 (Sept.).

Dezalay, Y. and Garth, B. (1998). Le 'Washington Consensus' – contribution à une sociologie de l'hégémonie du néoliberalisme, *Actes de la Recherche en Sciences Sociales* 121–2 (Mars): 3–20.

Dicken, P. (1999). Global Shift – The Role of United States Transnational Corporations. In Slater, D. and Taylor, P. J., eds, *The American Century*, Blackwell, Oxford, pp. 35–50.

Dietz, G. (1995). Zapatismo y Movimientos Étnicos-Regionales en México, *Nueva Sociedad* 140 (Nov.–Dic.): 33–50.

Diouf, J. (2002). Acabar con el Hambre, *Le Monde Diplomatique*, Edición Española, año VI, no. 80 (Junio): 9.

Dirlik, A. (1994). The Postcolonial Aura: Third World Criticism in the Age of Global Capitalism, *Critical Inquiry* 20 (Winter): 328–56.

Doornbos, M. (2001). 'Good Governance': The Rise and Decline of a Policy Metaphor? *Journal of Development Studies* 37(6) (Aug.): 93–108.

Dos Santos, T. (1998). The Theoretical Foundations of the Cardoso Government: A New Stage of the Dependency-theory Debate, *Latin American Perspectives* 25(1): 53–70.

Doty, R. (1996). *Imperial Encounters*, University of Minnesota Press, Minneapolis.

Doyle, M. W. (1986). *Empires*, Cornell University Press, Ithaca.

Drake, P. W. (1994)., The Political Economy of Foreign Advisers and Lenders in Latin America. In Drake, P. W., ed., *Money Doctors, Foreign Debts and Economic Reforms in Latin America from the 1890s to the Present*, Scholarly Resources Inc., Washington, pp. xi–xxxiii.

Duffield, M. (2001). *Global Governance and the New Wars*, Zed Books, London and New York.

Dunkerley, J. (1990). *Political Transition and Economic Stabilisation: Bolivia, 1982–1989*, Research Papers no. 22, Institute of Latin American Studies, University of London.

Dussel, E. (1998). *The Underside of Modernity*, Humanity Books, New York.

ECLAC (1989). *Preliminary Overview of the Economy of Latin America and the Caribbean*, Santiago, Chile.

ECLAC(1991). *Sustainable Development: Changing Production Patterns, Social Equity and the Environment*, Santiago, Chile.

Eisenstadt, S. N. (1966). *Modernization – Protest and Change*, Prentice Hall, New Jersey.

Eisenstadt, S. N. (2000). Multiple Modernities, *Daedalus*129(1) (Winter): 1–29.

Elguea, J. A. (1991). El Sangriento Camino hacia la Utopía: Las Guerras de Desarrollo en América Latina 1945–1989, *Estudios Sociológicos* 9(25): 145–64.

Escobar, A. (1995). *Encountering Development*, Princeton University Press, Princeton, NJ.

Etzold, T. H. and Gaddis J. L., eds (1978). *Containment: Documents on American Policy and Strategy, 1945–1950*, Columbia University Press, New York.

Evers, T. (1985). Identity: The Hidden Side of New Social Movements in Latin America. In Slater, D. ed., *The New Social Movements and the State in Latin America*, CEDLA, Amsterdam, pp. 43–71.

EZLN (1996). *Crónicas Intergalácticas – Primer Encuentro Intercontinental por la Humanidad y contra el Neoliberalismo*, Chiapas, Mexico.

Fabian, J. (1983). *Time and the Other*, Columbia University Press, New York.

Falk, R. (1993). The Making of Global Citizenship. In Brecher, J., et al., eds, *Global Visions – Beyond the New World Order*, Black Rose Books, Montreal, pp. 39–50.

Falk, R. (1999). *Predatory Globalization – A Critique*, Polity Press, Cambridge.

Fals Borda, O. (2002). El Tercer Mundo y la Reorientación de las Ciencias Contemporáneas, *Nueva Sociedad* 180–1: 169–81. Originally published in 1990.

Fanon, F. (1969). *The Wretched of the Earth*, Penguin Books, Harmondsworth.

Ferguson, J. (1995). From African Socialism to Scientific Capitalism: Reflections on the Legitimation Crisis in IMF-ruled Africa. In Moore, D. B. and Schmitz, G. J., eds, *Debating Development Discourse*, Macmillan, London, pp. 129–48.

Fine, B. (1999). A Question of Economics: Is it Colonizing the Social Sciences? *Economy and Society* 28(3) (Aug.): 403–25.

Fine, B. (2001). *Social Capital versus Social Theory*, Routledge, London and New York.

Fisher, J. (1995). Editorial: Some Thoughts on 'Contaminations', *Third Text* 32 (Autumn): 3–7.

Fisher, W. F. and Ponniah, T., eds (2003). *Another World is Possible*, Zed Books, London.

Foner, P. S. (1972). *The Spanish-Cuban-American War and the Birth of American Imperialism, vol II: 1898–1902*, Monthly Review Press, New York and London.

Fontana, B. (1993). *Hegemony and Power*, University of Minnesota Press, Minneapolis.

Foucault, M. (1979). *Discipline and Punish*, Peregrine Books, Harmondsworth.

Foucault, M. (1980a). *Power/Knowledge*, ed Colin Gordon. Pantheon Books, New York.

Foucault, M. (1980b). *The History of Sexuality, vol. 1: An Introduction*, Vintage Books, New York.

Foucault, M. (1984). Nietzsche, Genealogy, History. In Rabinow, P., ed., *The Foucault Reader*, Pantheon Books, New York, pp. 76–100.

Foucault, M. (1986). The Subject and Power: An Afterword. In Dreyfus, H. L. and Rabinow, P., eds, *Michel Foucault: Beyond Structuralism and Hermeneutics*, Harvester Press, Brighton, pp. 208–26.

Frank, A. G. (1967). *Capitalism and Underdevelopment in Latin America*, Monthly Review Press, New York.

Frank, A. G. (1969). *Latin America: Underdevelopment or Revolution*, Monthly Review Press, New York.

Fukuyama, F. (1992). *The End of History and the Last Man*, Penguin Books, London.

Furedi, F. (1994). *The New Ideology of Imperialism*, Pluto Press, London and Boulder.

Furtado, C. (1964). *Development and Underdevelopment*, University of California Press, Berkeley and Los Angeles. (Portuguese edition 1961.)

Furtado, C. (1969). US Hegemony and the Future of Latin America. In Horowitz, I. L., De Castro, J. and Gerassi, J., eds, *Latin American Radicalism*, Vintage Books, New York, pp. 61–74.

Gaddis, J. L. (1982)., *Strategies of Containment*, Oxford University Press, Oxford and New York.

Gane, M. (1990). Ironies of Postmodernism: Fate of Baudrillard's Fatalism, *Economy and Society* 19(3): 314–33.

Galeano, E. (1973). *Open Veins of Latin America*, Monthly Review Press, New York and London.

Galeano, E. (1983). *Days and Nights of Love and War*, Pluto Press, London.

Gamarra, E. A. (1994). *Entre la Droga y la Democracia*, ILDIS, La Paz.

Gamarra, E. A. (1999). The United States and Bolivia: Fighting the Drug War. In Bulmer-Thomas, V. and Dunkerley, J., eds, *The United States and Latin America: The New Agenda*, The Institute of Latin American Studies, University of London, and Harvard University Press, London and Cambridge, MA, pp. 177–206.

Gantenbein, J. W., ed. (1950). *The Evolution of Our Latin American Policy – A Documentary Record*, Columbia University Press, New York.

García-Canclini, N. (1991). Los Estudios Culturales de los 80 a los 90: perspectivas antropológicas y sociológicas en América Latina, *Iztapalapa: Revista de Ciencias Sociales y Humanidades* 11(24): 9–26.

García-Canclini, N. (1995). *Hybrid Cultures*, University of Minnesota Press, Minneapolis.

Gardner, L. C., LaFeber, W. F. and McCormick, T. J., eds (1973). *Creation of the American Empire: US Diplomatic History*, Rand McNally & Co., Chicago and London.

Gendzier, I. L. (1985). *Managing Political Change*, Westview Press, Boulder.

George, S. and Sabelli, F. (1994). *Faith and Credit – The World Bank's Secular Empire*, Penguin Books, London.

Gibbon, P. (1993). The World Bank and the New Politics of Aid. In Sørensen, G., ed., *Political Conditionality*, Frank Cass, London, pp. 35–62.

Giddens, A. (1999). *Runaway World*, Profile Books, London.

Giddens, A. (2000). *The Third Way and its Critics*, Polity Press, Cambridge.

Glassman, J. (2001). From Seattle (and Ubon) to Bangkok: The Scales of Resistance to Corporate Globalization, *Environment and Planning D: Society and Space* 19: 513–33.

Gleijeses, P. (1999). Afterword: The Culture of Fear. In Cullather, N., *The Secret History*, Stanford University Press, Stanford, pp. xix–xxxii.

Gómez, L. (1988). Deconstrucción o Nueva Síntesis: aproximación crítica a la noción de postmodernidad. In Calderón, F., ed., *Imagenes Desconocidas: La*

Modernidad en la Encrucijada Postmoderna, CLACSO, Buenos Aires, pp. 85–93.

Gómez-Peña, G. (1992). A Binational Performance Ritual, *Third Text* 19 (Summer): 64–78.

González-Casanova, P. (1969). Internal Colonialism and National Development. In Horowitz, I. L., De Castro, J. and Gerassi, J., eds, *Latin American Radicalism*, Vintage Books, New York, pp. 118–39.

González-Casanova, P., (1995a). *Colonialismo Global e a Democracia*, Civilização Brasileira, Rio de Janeiro.

González-Casanova, P. (1995b). Causes of the Rebellion in Chiapas, UNAM, Mexico City, unpublished paper.

González-Cruz, M. (1998). The US Invasion of Puerto Rico, *Latin American Perspectives*, 25(5) (Sept.): 7–26.

Gordon, C. (1991). Governmental Rationality: An Introduction. In Burchell, G., Gordon, C. and Miller, P., eds, *The Foucault Effect: Studies in Governmentality*, Harvester/Wheatsheaf, London and Toronto, pp. 1–51.

Government Printing Office (1943). Senate Document (188 of the 77th Congress, 2d Session), *Addresses and Messages of Franklin D. Roosevelt*, Washington DC (reprinted by His Majesty's Stationary Office, London, 1943).

Graebner, N., ed. (1968). *Manifest Destiny*, Bobbs-Merrill, New York.

Graham, C. (1992). The Politics of Protecting the Poor during Adjustment: Bolivia's Emergency Social Fund, *World Development* 20(9): 1233–51.

Gramsci, A. (1971). *Selections from the Prison Notebooks*, Lawrence & Wishart, London.

Gramsci, A. (1975). *The Modern Prince*, International Publishers, New York.

Gramsci, A. (1977). *Selections from Political Writings 1910–1920*, Lawrence and Wishart, London.

Graziano, F. (1992). *Divine Violence*, Westview Press, Boulder.

Green, D. (1995). *Silent Revolution –The Rise of Market Economics in Latin America*, Cassell, London.

Green, D. and Griffith, M. (2002). Globalization and its Discontents, *International Affairs* 78(1): 49–68.

Gregory, D. (1994). *Geographical Imaginations*, Blackwell, Oxford.

Griffin, K. B. (1969). *Underdevelopment in Spanish America*, George Allen and Unwin, London.

Gros, J.-G. (1996). Towards a Taxonomy of Failed States in the New World Order: Decaying Somalia, Liberia, Rwanda and Haiti, *Third World Quarterly* 17(3): 455–71.

Grovogui, S. N'Z. (1996). *Sovereigns, Quasi Sovereigns and Africans*, University of Minnesota Press, Minneapolis.

Guattari, F. (2000). *The Three Ecologies*, Athlone Press, London. (First published in France in 1989.)

Gülalp, H. (1998). The Eurocentrism of Dependency Theory and the Question of 'Authenticity': A View from Turkey, *Third World Quarterly* 19(5): 951–61.

Guyatt, N. (2000). *Another American Century?* Zed Books, London and New York.

Habermas, J. (1987). *The Philosophical Discourse of Modernity*, Polity Press, Cambridge.

Hahn, P. L. and Heiss M. A., eds (2001). *Empire and Revolution – The United States and the Third World since 1945*, Ohio State University Press, Columbus.

Hall, S. (1996). When was the 'Post-Colonial'? Thinking at the Limit. In Chambers, I. and Curtis, L., eds, *The Post-Colonial Question*, Routledge, London and New York, pp. 242–60.

Halpern, M. (1965). The Rate and Costs of Political Development, *Annals of the American Academy of Political and Social Sciences* 358 (March): 20–9.

Hammond, J. L. (1999). Law and Disorder: The Brazilian Landless Farmworkers' Movement, *Bulletin of Latin American Research* 18(4): 469–89.

Hardt, M. and Negri, T. (2000). *Empire*, Harvard University Press, Cambridge, MA, and London.

Harris, Nigel (1986). *The End of the Third World: Newly Industrializing Countries and the Decline of an Ideology*, Penguin Books, Harmondsworth.

Harvey, N. (1995). Rebellion in Chiapas: Rural Reforms and Popular Struggle, *Third World Quarterly* 16(1): 39–73.

Harvey, N. (1998). *The Chiapas Rebellion: The Struggle for Land and Democracy*, Duke University Press, Durham and London.

Hausmann, R. (1998). Fiscal Institutions for Decentralisng Democracies: Which Way to Go? In IDB-OECD, *Democracy, Decentralisation and Deficits in Latin America*, eds K. Fukasaku and R. Hausmann, Washington DC and Paris, pp. 13–32.

Healy, D. F. (1963). *The United States in Cuba 1898–1902*, University of Wisconsin Press, Madison.

Healy, D. F. (1970). *US Expansionism*, University of Wisconsin Press, Madison.

Hegel, G. W. F. (1956). *The Philosophy of History*, Dover Publications, New York. (First published in 1899.)

Hegel, G. W. F. (1967). *Philosophy of Right*, Oxford University Press, Oxford.

Held, D. (1995). *Democracy and the Global Order*, Polity Press, Cambridge.

Held, D. and McGrew, A. (2002). *Globalization/Anti-Globalization*, Polity Press, Cambridge.

Hemispheric Social Alliance (2001). *Alternatives for the Americas*, Discussion Draft no. 3, www.asc-hsa.org.

Hewitt de Alcántara, C. (1998). Uses and Abuses of the Concept of Governance, *International Social Science Journal* 155: 105–13.

Higgins, N. P. (2000). The Zapatista Uprising and the Poetics of Cultural Resistance, *Alternatives* 25(3) (July–Sept.): 359–74.

Higgins, N. P. (2001). Mexico's Stalled Peace Process: Prospects and Challenges, *International Affairs* 77(4): 885–903.

Higgott, R. A. (1983). *Political Development Theory*, Routledge, London and New York.

Hills, J. (1994). Dependency Theory and its Relevance Today: International Institutions in Telecommunications and Structural Power, *Review of International Studies* 20: 169–86.

Hinkelammert, F. J. (1999). *El Huracán de la Globalización*, Editorial Departamento Ecuménico de Investigaciones (DEI), San José, Costa Rica.

Hirschman, A. O. (1981). *The Passions and the Interests: Political Arguments for Capitalism Before its Triumph*, Princeton University Press, Princeton, NJ.

Hobsbawm, E. (1994). *Age of Extremes – The Short Twentieth Century 1914–1991*, Michael Joseph, London.

Holden, R. H. and Zolov, E., eds (2000). *Latin America and the United States – A Documentary Record*, Oxford University Press, Oxford and New York.

Honig, B. (1993). *Political Theory and the Displacement of Politics*, Cornell University Press, Ithaca, NY.

Hoogvelt, A. (1987). The Crime of Conditionality: An Open Letter to the IMF, *Review of African Political Economy* 38 (April): 80–5.

Hopenhayn, M. (1988). El Debate Post-Moderno y la Dimensión Cultural del Desarrollo. In Calderón, F., ed., *Imagenes Desconocidas: La Modernidad en la Encrucijada Postmoderna*, CLASCO, Buenos Aires.

Hopenhayn, M. (1995). Postmodernism and Neoliberalism in Latin America. In Beverley, J., Oviedo, J. and Aronna, M., eds, *The Postmodernism Debate in Latin America*, Duke University Press, Durham and London, pp. 93–109.

Hopenhayn, M. (2001). *No Apocalypse, No Integration*, Duke University Press, Durham and London.

Horsman, R. (1981). *Race and Manifest Destiny*, Harvard University Press, Cambridge, MA.

Hösle, V. (1992). The Third World as a Philosophical Problem, *Social Research* 59(2): 227–62.

Hunt, M. H. (1987). *Ideology and US Foreign Policy*, Yale University Press, New Haven and London.

Huntington, S. P. (1968). *Political Order in Changing Societies*, Yale University Press, New Haven and London.

Huntington, S. P. (1984). Will More Countries Become Democratic? *Political Science Quarterly* 99 (Summer): 193–218.

Huntington, S. P. (1998). *The Clash of Civilizations and the Remaking of World Order*, Touchstone Books, London and New York.

Husserl, E. (1965). *Phenomenology and the Crisis of Philosophy*, Harper and Row, New York.

Ianni, O. (1971). *Imperialismo y Cultura de la Violencia en América Latina*, Siglo XXI Editores, México DF.

Icaria Editorial (2002). *Porto Alegre (Foro Social Mundial 2002)*, Barcelona.

IDB (Inter-American Development Bank). (1991). *Economic and Social Progress in Latin America: 1991 Report*, Washington DC.

IDB (Inter-American Development Bank). (1994). *Economic and Social Progress in Latin America 1994 Report: Fiscal Decentralization*, Washington DC.

IDB (Inter-American Development Bank). (1998). *Economic and Social Progress in Latin America 1998–99 Report: Facing up to Inequality in Latin America*, Washington DC.

Ignatieff, M. (2003). *Empire Lite*, Vintage, London.

ILDIS/CEDLA (1994). *Informe Social Bolivia* no. 1, *Balance de Indicadores Sociales*, La Paz.

Immerman, R. (1982)., *The CIA in Guatemala – The Foreign Policy of Intervention*, University of Texas Press, Austin.

Isacson, A. (2001). Militarizing Latin American Policy, *Foreign Policy in Focus* 6(21) (May); www.fpif.org.

Jackson, R. H. (1993). *Quasi-States: Sovereignty, International Relations and the Third World*, Cambridge University Press, Cambridge.

Jalée, P. (1968). *The Pillage of the Third World*, Monthly Review Press, London and New York.

Jalée, P. (1969). *The Third World in World Economy*, Monthly Review Press, New York.

James, P. (1997). Postdependency? The Third World in an Era of Globalism and Late Capitalism, *Alternatives* 22(2) (April–June): 205–26.

Jameson, F. (1986). Foreword in J.-F. Lyotard, *The Postmodern Condition*, Manchester University Press, Manchester, pp. vii–xxi.

Jameson, F. (1992). *The Geopolitical Aesthetic*, Indiana University Press, Bloomington and Indianapolis.

Jameson, F. (1997). Culture and Finance Capital, *Critical Inquiry* 24 (Autumn): 246–65.

Jenkins, R. (1997). Trade Liberalisation in Latin America: The Bolivian Case, *Bulletin of Latin American Research* 16(3): 307–25.

Joseph, G. M, Legrand, C. C. and Salvatore, R. D., eds (1998). *Close Encounters of Empire – Writing the Cultural History of US–Latin American Relations*, Duke University Press, Durham.

Joyce, E. (1999). Packaging Drugs: Certification and the Acquisition of Leverage. In Bulmer-Thomas, V. and Dunkerley, J., eds, *The United States and Latin America: The New Agenda*, Institute of Latin American Studies, University of London and Harvard University Press, London and Cambridge, MA, pp. 207–25.

Julien, I. and Mercer, K. (1988). Introduction: de margin and de centre, *Screen* (Autumn): 2–10.

Kagan, R. and Kristol, W. (2000). The Present Danger, *The National Interest* (59) (Spring): 57–69.

Kaplan, A. (2002). *The Anarchy of Empire in the Making of US Culture*, Harvard University Press, Cambridge, MA, and London.

Kaplan, A. and Pease, D. E., eds (1993). *Cultures of United States Imperialism*, Duke University Press, Durham and London.

Kapur, G. (1997). Globalisation and Culture, *Third Text* 39 (Summer): 21–38.

Karney, S. (1989). *In Our Image*, Random House, New York.

Kay, C. (1989). *Latin American Theories of Development and Underdevelopment*, Routledge, London and New York.

Kay, C. and Gwynne, R. N. (2000). Relevance of Structuralist and Dependency Theories in the Neoliberal Period: A Latin American Perspective, *Journal of Developing Societies* 16(1): 49–69.

Keith, N. W. (1997). *Reframing International Development – Globalism, Postmodernity and Difference*, Sage Publications, Thousand Oaks, London and New Delhi.

Kenworthy, E. (1995). *America/Américas*, Pennsylvania State University Press, Pennsylvania.

King, A. D. (2003). Cultures and Spaces of Postcolonial Knowledges. In Anderson, K., et al., eds., *Handbook of Cultural Geography*, Sage, London, pp. 381–97.

Klare, M. T. (2001). The New Geography of Conflict, *Foreign Affairs* (May/June): 49–61.

Klare, M. T. (2002). *Resource Wars*, Henry Holt & Co., Owl Books, New York.

Klein, B. S. (1994). *Strategic Studies and World Order*, Cambridge University Press, Cambridge.

Kolko, G. (1988). *Confronting the Third World: United States Foreign Policy 1945–1980*, Pantheon Books, New York.

Kolko, G. (1997). *Vietnam – Anatomy of a Peace*, Routledge, London and New York.

Laclau, E. (1985). New Social Movements and the Plurality of the Social. In Slater, D., ed., *New Social Movements and the State in Latin America*, CEDLA, Amsterdam, pp. 27–42.

Laclau, E. (2000). Constructing Universality. In Butler, J. , Laclau, E. and Žižek, S., eds, *Contingency, Hegemony, Universality*, Verso Books, London and New York, pp. 281–307.

Laclau, E. and Mouffe, C. (2001). *Hegemony and Socialist Strategy*, Verso Books, London and New York.

LaFeber, W. (1963). *The New Empire*, Cornell University Press, Ithaca, NY.

LaFeber, W. (1989). *The American Age: United States Foreign Policy at Home and Abroad since 1750*, Norton & Co., New York and London.

Landell-Mills, P. (1992). Governance, Cultural Change and Empowerment, *Journal of Modern African Studies* 30(4): 543–67.

Larrain, J. (1989). *Theories of Development*, Polity Press, Cambridge.

Lasch, C. (1973). The Anti-Imperialist as Racist. In Paterson, T. G., ed., *American Imperialism and Anti-Imperialism*, Thomas Y. Crowell, New York, pp. 110–17.

Laski, H. J. (1946). *Reflections on the Revolution of our Time*, George Allen & Unwin, London.

Latham, M. E. (2000). *Modernization as Ideology*, University of North Carolina Press, Chapel Hill and London.

Latour, B. (1993). *We Have Never Been Modern*, Harvester Wheatsheaf, New York and London.

Lazreg, M. (1994). *The Eloquence of Silence: Algerian Women in Question*, Routledge, London and New York.

Lazreg, M. (2002). Development: Feminist Theory's Cul-de-Sac. In Saunders, K., ed., *Feminist Post-Development Thought*, Zed Books, London and New York, pp. 123–45.

Lechner, N. (1991). La Democratización en el Contexto de la Cultura Posmoderna, *Revista Foro* 14 (Abril): 63–70.

Leftwich, A. (1994). Governance, the State and the Politics of Development, *Development and Change* 25: 363–86.

Lehmann, D. (1990). *Democracy and Development in Latin America*, Polity Press, Cambridge.

Lerner, D. (1958). *The Passing of Traditional Society: Modernizing the Middle East*, Free Press, New York.

Lerner, M. (1964). The Triumph of Laissez Faire. In Schlesinger Jr, A. M. and White, M., eds, *Paths of American Thought*, Chatto & Windus, London, pp. 147–66.

Levy, M. J. (1952). Some Sources of the Vulnerability of the Structures of Relatively Nonindustrialized Societies to Those of Highly Industrialized Societies. In Hoselitz, B. F., ed., *The Progress of Underdeveloped Areas*, University of Chicago Press, Chicago & London, pp. 113–25.

Levy, M. J. (1966). *Modernization and the Structure of Societies – A Setting for International Affairs*, 2 vols, Princeton University Press, Princeton, NJ.

Lipietz, A. (1987). *Mirages and Miracles: The Crises of Global Fordism*, Verso, London.

Lipovetsky, G. (1994). *El Crepúsculo del Deber*, Editorial Anagrama, Barcelona.

Lucas, S. (1996). Campaigns of Truth: The Psychological Strategy Board and American Ideology, 1951–1953, *The International History Review* 18(2) (May): 279–301.

Luce, H. R. (1941). *The American Century*, Farrar & Rinehart, New York and Toronto.

Lustig, N. (1994). Medición de la Pobreza y de la Desigualdad en la América Latina: el Emperador no Tiene Ropa, *El Trimestre Económico* 61(1) (Enero–Marzo): 200–16.

Lutz, C. (1997). The Psychological Ethic and the Spirit of Containment, *Public Culture* 9: 135–59.

Lyotard, J.-F. (1986). *The Postmodern Condition – A Report on Knowledge*, Manchester University Press, Manchester.

Lyotard, J.-F. (1988). *The Differend – Phrases in Dispute*, Manchester University Press, Manchester.

Lyotard, J.-F. (1997). *Postmodern Fables*, University of Minnesota Press, Minneapolis.

MacPherson, C. B. (1988). *The Political Theory of Possessive Individualism – Hobbes to Locke*, Oxford University Press, Oxford.

Maidment, R. and Dawson, M., eds (1994). *The United States in the Twentieth Century – Key Documents*, Open University, Hodder & Stoughton, London.

Main, L. (2001). The Global Information Infrastructure: Empowerment or Imperialism? *Third World Quarterly* 22(1): 83–97.

Mainwaring, S. and Viola, E. (1984). New Social Movements, Political Culture and Democracy: Brazil and Argentina in the 1980s, *Telos* 61 (Fall): 17–54.

Mamdani, M. (1994). A Critical Analysis of the IMF Programme in Uganda. In Himmelstrand, U., Kinyanjui, K. and Mburugu, E., eds, *African Perspectives on Development*, James Curry Ltd, London, pp. 128–36.

Marini, R.M. (1975). *Subdesarrollo y Revolución*, Siglo XXI Editores, México DF.

Martí, J. (1961). *Obras Completas XIX: Estados Unidos y América Latina*, Patronato del Libro Popular, Havana.

Marx, K. and Engels, F. (1998). *The Communist Manifesto – A Modern Edition*, Verso Books, London and New York.

Mattelart, A. (1994). *Mapping World Communication*, University of Minnesota Press, Minneapolis and London.

May, E. R. (1973). *Imperial Democracy*, Harper & Row, New York and London.

Mazower, M. (1999). *Dark Continent*, Penguin, London.

McCarthy, T. (2002). *Vergangenheitsbewältigung* in the USA – On the Politics of the Memory of Slavery, *Political Theory* 30(5) (Oct.): 623–48.

McClelland, D. (1961). *The Achieving Society*, Von Nostrand, Princeton, NJ.

McDougall, W. A. (1997). *Promised Land, Crusader State*, Houghton Mifflin Co., Boston and New York.

McEwan, C. (2003). The West and Other Feminisms. In Anderson, K., et al., eds, *Handbook of Cultural Geography*, Sage, London, pp. 405–19.

McGrew, A. (1994). Introduction in A. McGrew, ed., *Empire – The United States in the Twentieth Century*, Hodder & Stoughton together with the Open University, London, pp. ix–xii.

Melucci, A. (1989). *Nomads of the Present*, Hutchinson, London.

Melucci, A. (1992). Liberation or Meaning: Social Movements, Culture and Democracy. In Nederveen Pieterse, J., ed., *Emancipations, Modern and Postmodern*, Sage, London, pp. 43–77.

Melucci, A. (1996). *Challenging Codes*, Cambridge University Press, Cambridge.

Mertes, T. (2002). Grass-roots Globalism, *New Left Review* 17 (Sept./Oct.): 101–10.

Meszaros, G. (2000). No Ordinary Revolution: Brazil's Landless Workers' Movement, *Race & Class* 42(2): 1–18.

Mignolo, W. D. (1995). *The Darker Side of the Renaissance*, University of Michigan Press, Ann Arbor.

Mignolo, W. D. (2000a). Human Understanding and (Latin) American Interests – The Politics and Sensibilities of Geohistorical Locations. In Schwarz, H. and Ray, S., eds., *A Companion to Postcolonial Studies*, Blackwell, Oxford, pp. 180–202.

Mignolo, W. D. (2000b). The Many Faces of Cosmo-Polis: Border Thinking and Critical Cosmopolitanism, *Public Culture* 12(3): 721–48.

Mignolo, W. D. (2000c). *Local Histories/Global Designs*, Princeton University Press, Princeton, NJ.

Mills, C. Wright (1956). *The Power Elite*, Oxford University Press, London and New York.

Miyoshi, M. (1993). A Borderless World? From Colonialism to Transnationalism and the Decline of the Nation State, *Critical Inquiry* 19 (Summer): 726–51.

Mohanty, C. T. (1988). Under Western Eyes: Feminist Scholarship and Colonial Discourses, *Feminist Review* 30 (Autumn): 61–88.

Mohanty, C. T. (1992). Feminist Encounters: Locating the Politics of Experience. In Barrett, M. and Phillips, A., eds, *Destabilizing Theory – Contemporary Feminist Debates*, Polity Press, Cambridge, pp. 74–92.

Moisy, C. (1997). Myths of the Global Information Village, *Foreign Policy* (Summer): 78–87.

Monereo, M. (2002). Porto Alegre II: en transición, *El Viejo Topo* 163: 12–14, Barcelona.

Moore, D. (1996). Reading Americans on Democracy in Africa: From the CIA to 'Good Governance', *The European Journal of Development Research* 8(1) (June): 123–48.

Moore, W. E. (1966). *Social Change*, Prentice Hall, New Jersey.

Moro, B. (2002). Una Recolonización Llamada 'Plan Puebla Panamá', *Le Monde Diplomatique*, Edición Española, año VII, no. 86 (dic.): 6–7.

Mosquera, G. (1994). Some Problems in Transcultural Curating. In Fisher, J., ed., *Global Visions: Towards a New Internationalism in the Visual Arts*, Kala Press, London, pp. 133–9.

Mouffe, C. (1988). Hegemony and New Political Subjects: Toward a New Concept of Democracy. In Nelson, C. and Grossberg, L., eds, *Marxism and the Interpretation of Culture*, University of Illinois Press, Urbana-Champaign, pp. 89–101.

Mouffe, C. (1995). Post-Marxism: Democracy and Identity, *Environment and Planning D: Society and Space* 13: 259–65.

Mouffe, C. (2000). *The Democratic Paradox*, Verso Books, London and New York.

Munck, R. (2000). *Marx@2000*, Zed Books, London and New York.

Muñoz, H. (2001). Good-bye USA? In Tulchin, J. S. and Espach, R. H., eds., *Latin America in the New International System*, Lynne Rienner Publishers, Boulder and London, pp. 73–90.

Muñoz, J. (2001). Rural Poverty and Development. In Crabtree, J. and Whitehead, L., eds, *Towards Democratic Viability: The Bolivian Experience*, Palgrave, Basingstoke, pp. 83–99.

Murphy, P. (1991). Postmodern Perspectives and Justice, *Thesis Eleven* (30): 117–32.

Myers, Gen. R. B. (2001). *Remarks at the Military Communications Conference*, MILCOM 2001, 31 Oct., Tysons Corner; www.dtic.mil/jcs/chairman/Milcom.htm, accessed 12 Feb. 2002.

Myers, Gen. R. B. (2002). *Global War on Terrorism*, www.dtic.mil/jcs/chairman/Posture_Statement.html, accessed 12 Feb. 2002.

Nadel, A. (1995). *Containment Culture*, Duke University Press, Durham and London.

Nakarada, R. (1994). Report from Zimbabwe, quoted in Editor's Introduction, *Alternatives*, 19(2) (Spring): 141–3.

Nandy, A. (1992). *The Intimate Enemy*, Oxford University Press, Delhi.

Nederveen Pieterse, J. (1991). Dilemmas of Development Discourse: The Crisis of Developmentalism and the Comparative Method, *Development and Change* 22: 5–29.

Nevins, J. (2002). *Operation Gatekeeper*, Routledge, London and New York.

Niess, F. (1990). *A Hemisphere to Itself: A History of US–Latin American Relations*, Zed Books, London and New Jersey.

Nietzsche, F. (1983). *Untimely Meditations*, Cambridge University Press, Cambridge and New York.

Nordstrom, C. (2000). Shadows and Sovereigns, *Theory, Culture & Society* 17(4) (Aug.): 35–54.

Nye Jr, J. S. (2002). *The Paradox of American Power*, Oxford University Press, Oxford and New York.

O'Brien, C. C. (1979). Modernisation, Order and the Erosion of a Democratic Ideal. In Lehmann, D., ed., *Development Theory*, Frank Cass, London, pp. 49–76.

O'Brien, T. F. (1996). *The Revolutionary Mission – American Enterprise in Latin America, 1900–1945*, Cambridge University Press, Cambridge.

OECD (1992). *Development Cooperation: 1992 Report*, Paris.

OECD (1994). *Development Cooperation: 1994 Report*, Paris.

OECD (1996). *Shaping the 21st Century: The Contribution of Development Co-operation*, May, Paris.

O'Gorman, E. (1972)., *The Invention of America*, Greenwood Press, Westport, CT.

Ong, A. (2000). Graduated Sovereignty in South-east Asia, *Theory, Society & Culture* 17(4) (Aug.): 55–75.

Oporto, H. (1991). *La Revolución Democrática – una nueva manera de pensar Bolivia*, Los Amigos del Libro, La Paz.

Ortiz, F. (1995). *Cuban Counterpoint*, Duke University Press, Durham and London.

Osorio, J. (1996). Actualidad de la Reflexión sobre el Subdesarrollo y la Dependencia: una visión crítica. In Marini, R. M. and Millán, M, eds, *La Teoría Social Latinoamericana*, Tomo IV, Ediciones El Caballito, Mexico, pp. 25–46.

Otto Wolf, F. (1986). Eco-socialist Transition on the Threshold of the 21st Century, *New Left Review* 158 (July/Aug.): 32–42.

Ó'Tuathail, G. (1998). Postmodern Geopolitics? The Modern Geopolitical Imagination and Beyond. In Ó'Tuathail, G. and Dalby, S., eds, *Rethinking Geopolitics*, Routledge, London and New York, pp. 16–38.

Ó'Tuathail, G. (2000). The Postmodern Geopolitical Condition: States, Statecraft and Security at the Millennium, *Annals of the Association of American Geographers* 90(1): 166–78.

Ó'Tuathail, G. and Agnew, J. (1992). Geopolitics and Discourse: Practical Geopolitical Reasoning in American Foreign Policy, *Political Geography* 11: 190–204.

Ó'Tuathail, G. and Dalby, S., eds (1998). *Rethinking Geopolitics*, Routledge, London and New York.

Padilla, F. M. (1992). *The Gang as an American Enterprise*, Rutgers University Press, New Brunswick, NJ.

Palast, G. (2003). *The Best Democracy Money Can Buy*, Robinson, London.

Parry, B. (1994). Signs of Our Times: A Discussion of Homi Bhabha's *The Location of Culture*, *Third Text* 28/29 (Autumn/Winter): 5–24.

Parsons, T. (1971). *The System of Modern Societies*, Prentice Hall, New Jersey.

Pasquino, P. (1993). Political Theory of War and Peace: Foucault and the History of Modern Political Theory, *Economy and Society* 22(1) (Feb.): 77–88.

Patterson, T. C. (1997). *Inventing Western Civilization*, Monthy Review Press, New York

Payer, C. (1991). *Lent and Lost – Foreign Capital and Third World Development*, Zed Books, London.

Pearce, F. (1978). *Crimes of the Powerful*, Pluto Press, London.

Peet, R., with Elaine Hartwick. (1999). *Theories of Development*, Guildford Press, New York.

Pérez Jr, L. A. (1997). *Cuba and the United States*, University of Georgia Press, Athens and London.

Perry Curtis Jr, L. (1997). *Apes and Angels: The Irishman in Victorian Caricature*, Smithsonian Institution Press, Washington.

Petras, J. (2002). The Imperial Counter-offensive: Contradictions, Challenges and Opportunities, *Journal of Contemporary Asia* 32(3): 291–303.

Philpott, D. (1995). Sovereignty: An Introduction and Brief History, *Journal of International Affairs*, 48(2) (Winter): 353–68.

Piscitelli, A. (1988). Sur, Post-Modernidad y Después. In Calderón, F., ed., *Imagenes Desconocidas: La Modernidad en la Encrucijada Postmoderna*, CLACSO, Buenos Aires, pp. 69–83.

Pletsch, C. E. (1981). The Three Worlds, or the Division of Social Scientific Labor, circa 1950–1975, *Comparative Studies in Society and History* 23(4): 565–90.

Polanyi, K. (1957). *The Great Transformation – The Political and Economic Origins of Our Time*, Beacon Press, Boston.

Portantiero, J. (1992). Foundations of a New Politics, *NACLA Report on the Americas* 25(5): 17–20.

Prakash, G. (1999). *Another Reason – Science and the Imagination of Modern India*, Princeton University Press, Princeton, NJ.

Pratt, J. W. (1927). The Origin of 'Manifest Destiny', *American Historical Review* 32 (July): 795–8.

Putnam, R. D. (1993). *Making Democracy Work*, Princeton University Press, Princeton, NJ.

Quijano, A. (1968). Tendencies in Peruvian Development and Class Structure. In Petras, J. and Zeitlin, M., eds, *Latin America: Reform or Revolution?* Fawcett, New York, pp. 289–328.

Quijano, A. (1988). Modernidad, Identidad y Utopía en América Latina. In Calderón, F., ed., *Imagenes Desconocidas: La Modernidad en la Encrucijada Postmoderna*, CLACSO, Buenos Aires, pp. 17–24

Quijano, A. (2000). Coloniality of Power, Eurocentrism and Latin America, *Nepantla* 1(3): 533–80.

Rabinow, P. (1986). Representations are Social Facts: Modernity and Postmodernity in Anthropology. In Clifford, J. and Marcus, G. E., eds, *Writing Culture: The Poetics and Politics of Ethnography*, University of California Press, Berkeley and London.

Ramonet, I. (1997). *Un Mundo sin Rumbo: crisis de fin de siglo*, Editorial Debate SA, Madrid.

Rancière, J. (1995). *On the Shores of Politics*, Verso, London and New York.

Reigadas, M. C. (1988). Neomodernidad y Posmodernidad: preguntado desde América Latina. In Mari, E., ed., *¿Posmodernidad?* Editorial Biblos, Buenos Aires, pp. 113–45.

Rich, P. (1997). NAFTA and Chiapas, *Annals of the American Academy of Political Science* 550 (March): 72–84.

Richard, N. (1987/88). Postmodernism and Periphery, *Third Text* (2): 5–12.

Richard, N. (1991). Latinoamérica y la Postmodernidad, *Revista de Crítica Cultural* 3 (Abril): 15–19.

Richard, N. (1992). Cultura, Política y Democracia, *Revista de Crítica Cultural* 5 (Julio): 5–7.

Richard, N. (1995). Cultural Peripheries: Latin America and Postmodernist Decentering. In Beverley, J., Oviedo, J. and Aronna, M., eds, *The Postmodernism Debate in Latin America*, Duke University Press, Durham and London, pp. 217–22.

Richardson, J. D. (1896). *A Compilation of the Messages and Papers of the Presidents*, vol. 1, Government Printing Office, Washington DC.

Richardson, J. D. (1905). *A Compilation of the Messages and Papers of the Presidents*, vol. X. Government Printing Office, Washington DC.

Rickard, J. (1998). The Occupation of Indigenous Space as 'Photograph'. In Barbican Art Gallery, *Native Nations: Journeys in American Photography*, ed. Jane Alison, London, pp. 57–71.

Rieff, D. (1991). *Los Angeles: Capital of the Third World*, Simon & Schuster, New York.

Rist, G. (1997). *The History of Development: From Western Origins to Global Faith*, Zed Books, London and New York.

Rivière, P. (2003). Patriot Act II, *Le Monde Diplomatique – edición española*, año VII, no. 89 (marzo): 12.

Robert, C. (2003). Emergencia de una Voz Africana, *Le Monde Diplomatique*, Edición Española, año VII, no. 88 (febrero): 28.

Robinson, J. (2003). Introduction to 'Beyond the West'. In Anderson, K., et al., eds, *Handbook to Cultural Geography*, Sage, London, pp. 399–404.

Robinson, W. I. (1996). *Promoting Polyarchy* , Cambridge University Press, Cambridge.

Robinson, W. I. and Norsworthy, K. (1985). Elections and US Intervention in Nicaragua, *Latin American Perspectives*, 12(2): 83–110.

Rocha, G. M. (2002). Neo-Dependency in Brazil, *New Left Review*, 16 (July/Aug.): 5–33.

Rodney, W. (1972). *How Europe Underdeveloped Africa*, Bogle-L'Ouverture Publications and Tanzania Publishing, London and Dar es Salaam.

Rodríguez Rabanal, C. (1989). *Cicatrices de la Pobreza: Un Estudio Psicoanalítico*, Editorial Nueva Sociedad, Caracas.

Rondinelli, D. A. (1981). Government Decentralization in Comparative Perspective: Theory and Practice in Developing Countries, *International Review of Administrative Sciences* 2: 133–45.

Roosevelt, T. (1889). *The Winning of the West*, vol. 1 of 4 vols, G. P. Putnam's Sons, New York.

Rorty, R. (1991a). *Objectivity, Relativism and Truth*, Philosophical Papers vol. 1, Cambridge University Press, Cambridge.

Rorty, R. (1991b). *Essays on Heidegger and Others*, Philosophical Papers vol. 2, Cambridge University Press, Cambridge.

Rorty, R. (1999). *Philosophy and Social Hope*, Penguin, Harmondsworth.

Rosenberg, E. S. (1982). *Spreading the American Dream*, Hill and Wang, New York.

Ross, D. (1992). *The Origins of American Social Science*, Cambridge University Press, Cambridge.

Rostow, W. W. (1960). *The Stages of Economic Growth: A Non-Communist Manifesto*, Cambridge University Press, Cambridge and New York.

Rostow, W. W. (1971). *Politics and the Stages of Growth*, Cambridge University Press, Cambridge.

Routledge, P. (2003). Convergence Space: Process Geographies of Grassroots Globalization Networks, *Transactions of the Institute of British Geographers* 28: 333–49.

Said, E. W. (1978). *Orientalism*, Penguin Books, London.

Said, E. W. (1981). *Covering Islam*, Routledge & Kegan Paul, London and Melbourne.

Said, E. W. (1989). Representing the Colonized: Anthropology's Interlocutors, *Critical Inquiry* 15: 205–25.

Said, E. W. (1993). *Culture and Imperialism*, Chatto and Windus, London.

Said, E. W. (2003). *Orientalism* (with a new preface). Penguin Books, London.

Sakai, N. (1997). *Translation & Subjectivity*, University of Minnesota Press, Minneapolis and London.

Salmi, J. (1993). *Violence and Democratic Society*, Zed Books, London and New Jersey.

Santos de Souza, B. (1999). On Oppositional Postmodernism. In Munck, R. and O'Hearn, D., eds, *Critical Development Theory*, Zed Books, London, pp. 29–43.

Santos de Souza, B. (2001). *Nuestra America*: Reinventing a Subaltern Paradigm of Recognition and Redistribution, *Theory, Culture and Society* 18(2–3): 185–217.

Sartre, J.-P. (1963). *The Search for a Method*, Vintage Books, New York.

Saunders, K. (2002). Introduction. In Saunders, K. ed., *Feminist Post-Development Thought*, Zed Books, London and New York, pp. 1–38.

Saxe-Fernández, J. (1969). The Central American Defense Council and Pax Americana: In Horowitz, I. L., De Castro, J. and Gerassi, J., eds, *Latin American Radicalism*, Vintage Books, New York, pp. 75–101.

Schatz, S. P. (1996). The World Bank's Fundamental Misconception in Africa, *The Journal of Modern African Studies* 34(2): 239–47.

Schmitz, G. J. (1995). Democratization and Demystification: Deconstructing 'Governance' as Development Paradigm. In Moore, D. B. and Schmitz, G. J., eds, *Debating Development Discourse*, Macmillan, London, pp. 54–90.

Schoultz, L. (1987). *National Security and United States Policy toward Latin America*, Princeton University Press, Princeton, NJ.

Schoultz, L. (1999). *Beneath the United States*, Harvard University Press, Cambridge, MA, and London.

Schumpeter, J. A. (1987). *Capitalism, Socialism and Democracy*, Unwin Paperbacks, London. (Originally published in 1943.)

Scott, C. V. (1995). *Gender and Development – Rethinking Modernization and Dependency Theory*, Lynne Rienner Publishers, Boulder and London.

Scott, J. M. (1996). *Deciding to Intervene – The Reagan Doctrine and American Foreign Policy*, Duke University Press, Durham.

Seoane, J. and Taddei, E. (2002). From Seattle to Porto Alegre: The Anti-neoliberal Globalization Movement, *Current Sociology* 50(1) (Jan.): 99–122.

Shapiro, M. J. (1997). *Violent Cartographies*, University of Minnesota Press, Minneapolis.

Sharp, J. (2003). Feminist and Postcolonial Engagements. In Agnew, J., Mitchell, K. and Toal, G., eds, *A Companion Guide to Political Geography*, Blackwell, Oxford, pp. 59–74.

Sherry, M. S. (1995). *In the Shadow of War*, Yale University Press, New Haven and London.

Shiva, V. (2003). Food Rights, Free Trade, and Fascism. In Gibney, M. J., ed., *Globalizing Rights*, Oxford University Press, Oxford, pp. 87–108.

Shohat, E. and Stam, R. (1994). *Unthinking Eurocentrism – Multiculturalism and the Media*, Routledge, London and New York.

Sidaway, J. (2000). Postcolonial Geographies: An Exploratory Essay, *Progress in Human Geography* 24(4): 591–612.

Sigmund, P. E. (1980). *Multinationals in Latin America*, University of Wisconsin Press, Madison and London.

Slater, D. (1974). Contribution to a Critique of Development Geography, *Canadian Journal of African Studies* 8(2): 325–54.

Slater, D. (1987). On Development Theory and the Warren Thesis: Arguments Against the Predominance of Economism, *Environment and Planning D: Society and Space* 5: 263–82.

Slater, D. (1989). *Territory and State Power in Latin America: The Peruvian Case*, Macmillan, London and New York.

Slater, D. (1992). Theories of Development and Politics of the Post-Modern, *Development and Change* 23(3): 283–319.

Slater, D. (1994). Exploring Other Zones of the Postmodern: Problems of Ethnocentrism and Difference across the North–South Divide. In Rattansi, A. and Westwood, S. eds, *Racism, Modernity & Identity*, Polity Press, Cambridge, pp. 87–125.

Slater, D. (1998). Post-Colonial Questions for Global Times, *Review of International Political Economy* 5(4) (Winter): 647–78.

Slater, D. (1999a). Situating Geopolitical Representations: Inside/Outside and the Power of Imperial Interventions. In Massey, D. and Allen, J., eds, *Human Geography Today*, Polity Press, Cambridge, pp. 62–84.

Slater, D. (1999b). Locating the American Century: Themes for a Post-Colonial Perspective. In Slater, D. and Taylor, P. J., eds, *The American Century*, Blackwell, Oxford, pp. 17–31.

Slater, D. (2002). Other Domains of Democratic Theory: Space, Power and the Politics of Democratization, *Environment and Planning D: Society and Space* 20: 255–76.

Slater, D. (2003a). Beyond Euro-Americanism: Democracy and Post-Colonialism. In Anderson, K., et al., eds, *Handbook of Cultural Geography*, Sage, London, pp. 420–32.

Slater, D. (2003b). Geopolitical Themes and Post-Modern Thought. In Agnew, J., et al., eds, *A Companion to Political Geography*, Blackwell, Oxford, pp. 75–91.

Sloterdijk, P. (1988). *Critique of Cynical Reason*, Verso Books, London and New York.

Slotkin, R. (1998). *The Gunfighter Nation*, University of Oklahoma Press, Norman.

Smith, A. (1980). *The Geopolitics of Information*, Faber & Faber, London and Boston.

Smith, P. H. (2000). *Talons of the Eagle: Dynamics of US–Latin American Relations*, Oxford University Press, New York and Oxford.

Smith, R. (1998). Closing the Door on Undocumented Workers, *NACLA Report on the Americas* 31(4): 6–9.

Smith, T. (1994). *America's Mission*, Princeton University Press, Princeton, NJ.

Soper, K. (1991). Postmodernism and its Discontents, *Feminist Review* 39: 97–108.

Spanos, W. V. (2000). *America's Shadow: An Anatomy of Empire*, University of Minnesota Press, Minneapolis.

Spivak, G. C. (1988). Can the Subaltern Speak? In Nelson, C. and Grossberg, L., eds, *Marxism and the Interpretation of Culture*, University of Illinois Press, Urbana, IL, pp. 271–313.

Spivak, G. C. (1990). *The Post-Colonial Critic*, ed. Sarah Harasym, Routledge, London and New York.

Spivak, G. C. (1999). *A Critique of Postcolonial Reason*, Harvard University Press, Cambridge, MA, and London.

Spurr, D. (1993). *The Rhetoric of Empire*, Duke University Press, Durham.

Stavenhagen, R. (1968). Seven Fallacies about Latin America. In Petras, J. and Zeitlin, M, eds, *Latin America: Reform or Revolution?* Fawcett, New York, pp. 13–31.

Stephanson, A. (1995). *Manifest Destiny*, Hill and Wang, New York.

Stephanson, A. (1998). Fourteen Notes on the Very Concept of the Cold War. In Ó'Tuathail, G. and Dalby, S., eds, *Rethinking Geopolitics*, Routledge, London and New York, pp. 62–85.

Strang, D. (1996). Contested Sovereignty: The Social Construction of Colonial Imperialism. In Biersteker, T. J. and Weber, C., eds, *State Sovereignty as Social Construct*, Cambridge University Press, Cambridge, pp. 22–49.

Sutcliffe, B. (2001). *100 Ways of Seeing an Unequal World*, Zed Books, London and New York.

Subcomandante Marcos (2000). El Fascismo Liberal, *Le Monde Diplomatique*, Edición Española, año V, nos 58–9 (Set.): 25–28.

Subcomandante Marcos (2001a). *Our Word is our Weapon, selected writings*, ed. J. Ponce de León, Seven Stories Press, New York.

Subcomandante Marcos (2001b). *Los del Color de la Tierra – textos insurgentes desde Chiapas*, Editorial Txalaparta, Tafalla, Spain.

Takaki, R. (1993). *A Different Mirror: A History of Multicultural America*, Back Bay Books, Boston and London.

Tandon, Y. (1994). Recolonization of Subject Peoples, *Alternatives* 19(2) (Spring): 173–83.

Taylor, J. G. (1979). *From Modernization to Modes of Production*, Macmillan, London.

Taylor, P. J. (1995). Beyond Containers: Inter-Nationality, Inter-Stateness, Inter-Territoriality, *Progress in Human Geography* 19: 1–15.

Taylor, P. J., Johnston, R. J. and Watts, M. (2002). Geography/Globalization. In Johnston, R. J., Taylor, P. J. and Watts, M, eds, *Geographies of Global Change*, Blackwell, Oxford, pp. 1–17.

Teivainen, T. (2002). The World Social Forum and Global Democratisation: Learning from Porto Alegre, *Third World Quarterly* 23(4): 621–32.

Thomas, C. (1999). Where is the Third World Now? *Review of International Studies* 25: 225–44.

Thomas, H. (2001). *Cuba*, Pan Books, London.

Tocqueville, Alexis de (1990). *Democracy in America*, Vintage, New York.

Tokman, V. E. (1994). Informalidad y Pobreza: Progreso Social y Moderniza-ción Productiva, *El Trimestre Económico* 61(1) no. 241 (Enero–Marzo): 177–99.

Torgovnick, M. (1991). *Gone Primitive*, University of Chicago Press, Chicago and London.

Torrico, E. R. (1990). Bolivia: el Rediseño Violento de la Sociedad Global, *Nueva Sociedad* 105: 153–63.

Trouillot, M.-R. (1995). *Silencing the Past – Power and the Production of History*, Beacon Press, Boston.

Turner, F. J. (1896). The Problem of the West, *The Atlantic Monthly* 78 (Sept.): 295–7.

Turner, F. J. (1962). *The Frontier in American History*, Holt, Rinehart and Winston, New York and London. (Originally published in 1920.)

UNDP (1998). *Human Development Report 1998*, Oxford University Press, New York and Oxford.

UNDP (2000). *Human Development Report 2000*, Oxford University Press, New York and Oxford.

UNDP (2002). *Human Development Report 2002*, Oxford University Press, New York and Oxford.

United States General Printing Office (1996). *Cuban Liberty and Democratic Solidarity (LIBERTAD). Act of 1996*, Public Law 104-111, 12 March 1996, 104th Congress, Washington DC.

Urbinati, N. (1998). From the Periphery of Modernity – Antonio Gramsci's Theory of Subordination and Hegemony, *Political Theory* 26(3) (June): 370–91.

Vattimo, G. (1991). *The End of Modernity*, Polity Press, Cambridge.

Vattimo, G. (1992). *The Transparent Society*, Polity Press, Cambridge.

Vázquez Montalbán, M. (1999). *Marcos: El Señor de los Espejos*, Grupo Santillana de Ediciones, Madrid and Aguilar, Buenos Aires.

Wainwright, H. (2003). Making a Peoples' Budget in Porto Alegre, *NACLA Report*, 36(5) (March/April): 37–42.

Walker, R. B. J. (1993). *Inside/Outside: International Relations as Political Theory*, Cambridge University Press, Cambridge.

Warren, B. (1980). *Imperialism: Pioneer of Capitalism*, New Left Review Editions, London.

Watkins, K. (1994). Debt Relief for Africa, *Review of African Political Economy* 62: 599–609.

Watnick, M. (1952). The Appeal of Communism to the Peoples of Underdeveloped Areas, *Economic Development and Cultural Change* 1: 22–36.

Watts, M. (1996). Islamic Modernities? Citizenship, Civil Society and Islamism in a Nigerian City, *Public Culture* 8: 251–89.

WDM (World Development Movement). (2003). *WDM in Action*, Summer, London, www.wdm.org.uk.

Weber, C. (1995). *Simulating Sovereignty*, Cambridge University Press, Cambridge.

Weber, C. (1999). *Faking It – US Hegemony in a 'Post-Phallic Era'*, University of Minnesota Press, Minneapolis.

Weber, M. (1978). *Economy and Society*, vol. 1, University of California Press, Berkeley.

Weber, M. (1992). *The Protestant Ethic and the Spirit of Capitalism*, Routledge, London and New York.

Weffort, F. C. (1991). La América Errada, *Revista Foro*, Bogotá, 15 (Sept.): 90–108.

Weinberg, A. (1963). *Manifest Destiny*, Quadrangle Paperbacks, Chicago.

Weldes, J. and Saco, D. (1996). Making State Action Possible: The United States and the Discursive Construction of 'The Cuban Problem', 1960–1994, *Millennium* 25(2) (Summer): 361–95.

Weston, R. (1972). *Racism in US Imperialism*, University of South Carolina Press, Columbia, South Carolina.

The White House (1997). *Clinton Memo on Drug War Certification*, Office of the Press Secretary, 28 Feb. 1997, Washington DC; usiswww/drugs3.html.

The White House (2002). *The National Security Strategy of the United States of America*, Sept., Washington DC.

Whyte, W. H. (1968). Imitation or Innovation: Reflections on the Institutional Development of Peru, *Administrative Science Quarterly* 13(13): 370–85.

Williams, D. E. (1990). Crisis and Renewal in the Social Sciences and the Colonies of Ourselves, *International Political Science Review* 11(1) (Jan.): 59–74.

Williams, P. (2003). One Nation under Suspicion, Loses Liberty and Justice for All, *The Times Higher*, 28 Feb., London: 18–19.

Williams, W. A. (1980). *Empire as a Way of Life*, Oxford University Press, Oxford.

Williamson, J. (1993). Democracy and the 'Washington Consensus', *World Development* 21(8): 1329–36.

Wilson, Gilmore R. (2002). Race and Globalization. In Johnston, R. J., Taylor, P. J. and Watts, M. J., eds, *Geographies of Global Change*, Blackwell, Oxford, pp. 261–74.

Wolfensohn, J. D. (1998). *The Other Crisis*, Address to the Board of Governors, World Bank, Washington DC.

Wolfowitz, P. (2000). Remembering the Future, *The National Interest* 59 (Summer): 35–45.

World Bank (1981a). *World Development Report 1981*, Oxford University Press, New York and Oxford.

World Bank (1981b). *World Bank Annual Report 1981*, Washington DC.

World Bank (1986). *World Development Report 1986*, Oxford University Press, New York and Oxford.

World Bank (1988). *World Development Report 1988*, Oxford University Press, New York and London.

World Bank (1991). *World Development Report 1991*, Oxford University Press, New York and London.

World Bank (1992a). *The World Bank Annual Report 1992*, Washington DC.

World Bank (1992b). *World Development Report 1992*, Oxford University Press, New York and Oxford.

World Bank (1992c). *Governance and Development*, Washington DC.

World Bank (1997). *World Development Report 1997*, Oxford University Press, New York and Oxford.

World Bank (1999). *World Development Report 1998/99*, Oxford University Press, New York.

World Bank (2000a). *World Development Report 1999/2000*, Oxford University Press, New York and Oxford.

World Bank (2000b). *The World Bank Annual Report 1999*, Oxford University Press, New York and Oxford.

World Bank (2003). *World Development Report 2003*, Oxford University Press, New York and Oxford.

Worsley, P. (1979). How Many Worlds? *Third World Quarterly* 1(2): 100–7.

Worsley, P. (1984). *The Three Worlds*, Weidenfeld and Nicolson, London.

Young, R. J. C. (1990). *White Mythologies*, Routledge, London and New York.

Young, R. J. C. (2001). *Postcolonialism*, Blackwell, Oxford.

Yúdice, G. (1996). El Impacto Cultural del Tratado de Libre Comercio Norteamericano. In García-Canclini, N., ed., *Culturas en Globalización*, Editorial Nueva Sociedad, Caracas, pp. 73–126.

Yúdice, G. (1998). The Globalization of Culture and the New Civil Society. In Alvarez, S., Dagnino, E. and Escobar, A., eds, *Cultures of Politics, Politics of Cultures*, Westview Press, Boulder and Oxford, pp. 353–79.

Zacher, M. W. (2001). The Territorial Integrity Norm: International Boundaries and the Use of Force, *International Organization* 55(2) (Spring): 215–50.

Zavala, I. M. (1988). On the (Mis-)Uses of the Post-Modern: Hispanic Modernism Revisited. In D'Haen, T. and Bertens, H., eds, *Postmodern Fiction in Europe and the Americas*, Editions Rodopi, Amsterdam and New York, pp. 83–113.

Zea, L. (1963). *The Latin American Mind*, University of Oklahoma Press, Norman.

Zea, L. (1970). *América en la Historia*, Ediciones Castilla SA, Madrid. (First published in 1957.)

Zea, L. (1989). *Simón Bolívar, integración en la libertad*, Monte Avila Editores Latinoamericana, Caracas, Venezuela.

Zelikow, P. (2003). The Transformation of National Security, *The National Interest* 71 (Spring): 17–28.

Zermeño, S. (1989). El Regreso del Líder: crisis, neoliberalismo y desorden, *Revista Mexicana de Sociología* 101 (Oct.–Dic.): 115–50.

Zermeño, S. (1995). Zapatismo, Región y Nación, *Nueva Sociedad*, 140, Noviembre–Diciembre, pp 51–57.

Zermeño, S. (1996). *La Sociedad Derrotada*, Siglo XXI Editores, Mexico.

Zinn, H. (1980). *A People's History of the United States*, Longman, London and New York.

Žižek, S. (1990). Beyond Discourse Analysis. In Laclau, E., ed., *New Reflections on the Revolution of Our Time*, Verso, London, pp. 249–60.

Žižek, S. (1998). A Leftist Plea for 'Eurocentrism', *Critical Inquiry* (Summer): 988–1009.

Žižek, S. (2000). Class Struggle or Postmodernism? Yes, Please! In Butler, J., Laclau, E. and Žižek, S., eds, *Contingency, Hegemony, Universality*, Verso, London and New York, pp. 90–135.

Index

NOTE: Page numbers followed by *n* indicate that information is in a note.